Diffusion Models of Environmental Transport

Diffusion Models of Environmental Transport

Bruce Choy, Ph.D.
Department of Chemical Engineering
The University of Sydney
Australia

Danny D. Reible, Ph.D.
Hazardous Substance Research Center (S&SW)
Department of Chemical Engineering
Louisiana State University
Baton Rouge

CRC Press
Taylor & Francis Group
Boca Raton London New York

CRC Press is an imprint of the
Taylor & Francis Group, an **informa** business

CRC Press
Taylor & Francis Group
6000 Broken Sound Parkway NW, Suite 300
Boca Raton, FL 33487-2742

First issued in paperback 2019

ISBN-13: 978-1-56670-414-4 (hbk)
ISBN-13: 978-0-367-39934-4 (pbk)
Library of Congress Card Number 99-41787

Library of Congress Cataloging-in-Publication Data

Diffusion models of environmental transport
edited by Bruce Choy and Danny Reible
p. cm.
Includes bibliographical references and index.
ISBN 1-56670-414-6 (alk. paper)
1. Environmental chemistry—Mathematical models. 2. Transport theory—Mathematical models. 3. Diffusion—Mathematical models.
I. Reible, Danny D. II. Title.
TD193.C487 1999
628.5—dc21 99-41787
CIP

Visit the Taylor & Francis Web site at
http://www.taylorandfrancis.com

and the CRC Press Web site at
http://www.crcpress.com

Preface

We are still trying to determine whether it is through the process of "diffusion" or simply by the overall action of "entropy" by which notes and calculations go missing over the years. I guess habits of doing calculations on backs of envelopes, margins in printouts, backs of photocopies, on little Post-it notes, or any scrap of paper around do not help the situation. Add to that the occasional haphazard filing and bookkeeping, and we've often found ourselves redoing calculations done last month or last year over and over again. Tired of this "natural" process we started to put together this document.

This document is meant to become a little, self-contained recipe book of models and evaluation procedures for a number of environmental transport calculations. It is hoped we can pass, to anyone with a scientific or engineering background, all the models in this publication so that they can be evaluated and applied. These models are ready to be picked up and used without further detailed mathematical manipulations. However, details of the derivation of these models have also been included for those who wish for a deeper understanding or who seek to add to the models presented.

This volume is dedicated to our families in the sincere hope that it will enable us to spend more rather than less time with them.

Cheers,

BC and DDR
October 1999

About the Authors

Bruce Choy

Dr. Choy is a lecturer in Chemical Engineering at the University of Sydney, Australia. During 1997 and 1998, he was a Research Associate with the Hazardous Substance Research Center/South and Southwest at Louisiana State University. In 1999, he received a Ph.D. on "Volatile Emissions of Organic Contaminants from Soils and Sediments" from the University of Sydney, under the supervision of Professor Danny Reible and Dr. Barry Walsh.

Danny D. Reible

Dr. Reible is Chevron Professor of Chemical Engineering at Louisiana State University and Director of the Hazardous Substance Research Center/South and Southwest, an EPA-supported consortium of Louisiana State University, Rice University, and Georgia Tech. He joined Louisiana State University after receiving a Ph.D. in Chemical Engineering in 1982 from the California Institute of Technology. From 1993 to 1995 he served as the Shell Professor of Environmental Engineering at the University of Sydney, Australia. In 1995 he returned to LSU after being named the Director of the Hazardous Substance Research Center and in 1998 he was named the Chevron Professor. He is also the author of the textbook, *Fundamentals of Environmental Engineering* and more than 100 technical papers and reports.

Table of Contents

List of figures

Nomenclature

c_A^{sat}	saturated vapor concentration of species A [ML^{-3}]
$c_{A,surf}$	mobile phase species A concentration at the surface [ML^{-3}]
c_A^{∞}	bulk vapor concentration of species A [ML^{-3}]
c_A^{air}	concentration of species A in the air phase per unit volume [ML^{-3}]
c_A^{water}	concentration of species A in the water phase per unit volume [ML^{-3}]
c_A^{soil}	concentration of species A in the soil phase per unit mass [M(A)M(soil)$^{-1}$]
c_A	mobile phase species A concentration [ML^{-3}]
$\mathscr{D}_A$	molecular diffusion coefficient of species A in the mobile phase [L^2T^{-1}]
$D_{A(eff)}$	effective diffusion coefficient of species A [L^2T^{-1}]
D_A	dispersion coefficient [L^2T^{-1}]
$D_{L(app)}$	apparent liquid diffusivity [L^2T^{-1}]
f_{oc}	fraction of organic carbon in soil [-]
$\Delta\hat{H}_{abs}$	molar heat of absorption of monolayer [$ML^2T^{-2}N^{-1}$]
$\Delta\hat{H}_{cond}$	molar heat of condensation of monolayer [$ML^2T^{-2}N^{-1}$]
H_c	air-water equilibrium partition coefficient (Henry's constant) [ML^{-3}(air)/ ML^{-3}(water)]
j_A	mass flux of component A [$ML^{-2}T^{-1}$]
k_1	first-order reaction rate constant [T^{-1}]
k_a	surface mass transfer coefficient [LT^{-1}]
$K_{air-water}$	air-water equilibrium partition coefficient [M L^{-3} (air)/ML^{-3}(water)]
$K_{oc-water}$	organic carbon-air equilibrium partition coefficient [MM^{-1}(oc)/ML^{-3}(water)]
$K_{octanol-water}$	octanol-water equilibrium partition coefficient [M L^{-3} (octanol)/ML^{-3}(water)]
$K_{soil-air}$	soil-air equilibrium partition coefficient [MM^{-1}(soil)/ML^{-3}(air)]
L_x	characteristic fetch length across the contaminated zone in the direction of the wind [L]
L_{char}	characteristic system length [L]

MW	molecular weight $[NM^{-1}]$
$\dot{n}_A$	mass flow of component A $[MT^{-1}]$
p_A	partial pressure of species A $[ML^{-1}T^{-2}]$
p_A^{sat}	saturated vapor pressure of species A $[ML^{-1}T^{-2}]$
R	Universal gas constant (8.314 J.mol^{-1}K^{-1}) $[ML^3T^{-1}N^{-1}\theta]$
R_f	retardation factor [-]
t	time [T]
T	absolute temperature $[\theta]$
u_x	average undisturbed air velocity above sediment surface (usually taken at a height of 10 m) $[LT^{-1}]$
v	Darcy velocity of mobile phase $[LT^{-1}]$
W	mass adsorbed onto a unit weight of soil $[MM^{-1}$(soil)]
W_m	mass of an adsorbed monolayer of molecules per unit weight of soil $[MM^{-1}$(soil)]
x_A	mole fraction of species A in the liquid phase [-]
$\bar{y}$	depth of clean cap [L]
y	depth from surface of the free liquid volatile organic [L]
z	depth or position [L]
β	Eigenvalue
ε	porosity of the medium [-]
Θ	moisture content
ρ_{air}	density of air $[ML^{-3}]$
ρ_A	mass concentration of liquid species A $[ML^{-3}]$
ρ_A	density of species A $[ML^{-3}]$
ρ_A^{field}	mass of free volatile organic liquid per unit volume of sediment $[ML^{-3}]$
ρ_{bulk}	bulk density of the porous medium $[ML^{-3}]$
τ	tortuosity of the medium [-]
μ_{air}	viscosity of air $[ML^{-1}T^{-1}]$
μ_A	viscosity of species A $[ML^{-1}T^{-1}]$
Ψ	Eigenfunction

Dimensionless groups

$$\mathrm{Re} = \frac{\rho \cdot u_x \cdot L_x}{\mu}$$

Reynolds number

$$\mathrm{Sc} = \frac{\mu_A}{\rho_A \cdot \mathscr{D}_A}$$

Schmidt number

$$Sh = \frac{k_a L_x}{\mathscr{D}_A}$$

Sherwood number

$$\mathrm{Pe} = \frac{v \cdot L_{char}}{D_A}$$

Peclet number

1 Environmental transport modeling

1.1 Introduction

Risk-based environmental decision making is crucial to responding effectively and efficiently to environmental problems. The identification of an environmental problem and the definition of an appropriate response in such a framework rely on having quantitative tools that relate the sources of the environmental problem to the human or ecological receptors that are negatively impacted. These quantitative tools include conceptual, physical, and mathematical models of the environmental and receptor processes that transport, dilute, degrade, and mitigate the effects of environmental contaminants. These fate and transport models define the exposure of receptors to environmental contaminants, which when combined with information on the adverse ecological effects of contaminants fully define their risks.

To be useful, these models must represent a good understanding and description of the key environmental processes responsible for the fate and transport of the environmental contaminants. Due to the complexity and multiplicity of the processes that influence environmental contaminants it is not possible to develop general models that quantitatively describe all processes and effects. It is necessary to balance the complexity of the models with the sometimes crude data and process knowledge available to support them. In addition, it is often convenient to employ fate and transport models to explore particular mechanisms or to define sensitivities to assumptions. As a result of these considerations, it is often convenient to employ analytical or semi-analytical solutions of the fate and transport equations as the basis for quantitative models.

Quantitative models must also recognize that almost all environmental processes are dynamic. The evaluation of the fate and transport of contaminants in the environment has often been limited to consideration of the equilibrium partitioning of these contaminants between environmental phases. While equilibrium partitioning is an important consideration and often influences both the final state of the environment and the dynamics leading to that final state, this concept rarely, if ever, provides an accurate model of environmental conditions. Regulatory or remedial efforts based upon equilibrium concepts are unlikely to be cost-effective and may, depending upon the situation, underestimate or overestimate the exposure and resulting effects of environmental contaminants.

The processes that must be included in models of environmental fate and transport processes of particular contaminants vary greatly. All contaminants, however, are subject to molecular diffusion, the mixing process associated with the random motion of molecules. In certain situations, advection, or transport by the bulk fluid motion, and reaction, for example by biodegradation or spontaneous degradation of contaminants, are also important. Other processes of importance in the environment such as turbulent mixing, dispersion in random porous media, and bioturbation of soil and sediments are often modeled as effective diffusion processes. Because a quantitative understanding of these processes is often lacking, lumping their effects into an effective diffusion coefficient is often convenient and more sophisticated approaches are sometimes simply not supported by the available data. It should be emphasized, however, that such diffusion coefficients are likely to be much greater in magnitude than molecular diffusion coefficients and are difficult, if not impossible, to extrapolate to other conditions.

The geometry of environmental fate and transport problems also varies greatly. Both the geometry of the domain of interest (large area or volume, regular or irregular) and the conditions imposed on the "boundary" of that domain vary. Broadly applicable environmental fate and transport models must exhibit the capability of addressing a wide variety of geometries and conditions.

Based upon the above discussion, useful models of fate and transport processes should possess at least the following characteristics:

- convenient to use and efficient to evaluate
- describe transient problems
- recognize the importance of partitioning into other fluid and solid phases
- describe diffusive or diffusive-like processes
- describe reaction and advective processes where necessary
- generally applicable to a variety of geometries and boundary conditions

The purpose of this document is to summarize a variety of analytical and semi-analytical material balance-based models that have these qualities. These include solutions of the diffusion and advection-diffusion equations with or without reaction in a variety of geometries and boundary conditions. The solutions describe the transient behavior of systems governed by these equations; in some cases transient transport coefficients, for example, diffusion coefficients, are considered. It is hoped that the solutions will provide a useful set of modeling tools that can be used to define fate, transport, and exposure from specific sources of environmental contaminants. In particular, the solutions allow the evaluation of transient diffusive, advective, or reactive processes on the distribution of contaminants in the environment. These solutions can be used to predict the effects of these processes or used in an inverse sense to determine effective transport or reaction coefficients that describe a set of environmental or laboratory data. The focus in this volume is on the solutions themselves and the discussion does not generally seek to develop models for particular transport problems. It is left to the reader to identify which of these solutions most appropriately models a given set of conditions in the environment or in a laboratory experiment.

Although the models are presented generally, the specific models, geometries, and boundary conditions that are included are directed at a variety of problems of interest to the authors in the evaluation of field problems or in the interpretation of laboratory data. These problems primarily include advection/diffusion/reaction problems in one-dimensional multilayered media. The applications of interest include transport and fate processes of contaminants crossing the sediment-water or soil-air interface in which vertical heterogeneity is more important than heterogeneities of either contaminant or media in areal directions. The surface layers of sediment and soils tend to be layered due to depositional patterns. As a result advective processes tend to be less dominant in the vertical than in the horizontal. Thus the vertical movement of contaminants through these layers is often controlled or strongly influenced by diffusion. In many situations, multiple layers of low permeability material may be placed specifically for the control of permeability and advective flow. In the laboratory, it is generally possible to design experiments to explore specific transport processes and diffusion-controlled experiments are often used to define effective diffusion, sorption, or reaction rates. It is to models of diffusion-dominated conditions in the environment or in laboratory experiments that this volume is dedicated. Some of the models have been reported elsewhere but have been included here for completeness. Some of the models have apparently not been reported in the open literature. The unique contributions of this volume are to models of diffusion through multiple layers of media with nonuniform initial conditions, differing transport and partitioning behavior between layers, or models of diffusion in layers of media subject to time-dependent transport or partitioning behavior. It is hoped that the present volume can serve as a useful single resource for such models that can complement the much broader collection of diffusion equation solutions provided in the classic reference texts by Crank, and Carslaw and Jaeger.

2 Preliminaries

2.1 Equilibrium between environmental phases

There exist a number of different phases in the natural environment (e.g. air, water, soil) in which a chemical contaminant may reside adsorbed on or dissolved within. Thermodynamic properties of the contaminant and these phases govern the potential distribution of the chemical species in the environment. Determining the preferred phase of a contaminant is the first stage in analyzing for the contaminant's potential for environmental impact. Note that the purpose of the models presented in this volume is ultimately to define rates of transport between phases. The equilibrium distribution is generally never observed in bulk phases in the environment. Equilibrium still defines the direction of transport (i.e., toward equilibrium) and, by virtue of an interface having effectively zero volume or accumulation, defines the ratio of concentrations in the immediate vicinity of an interface.

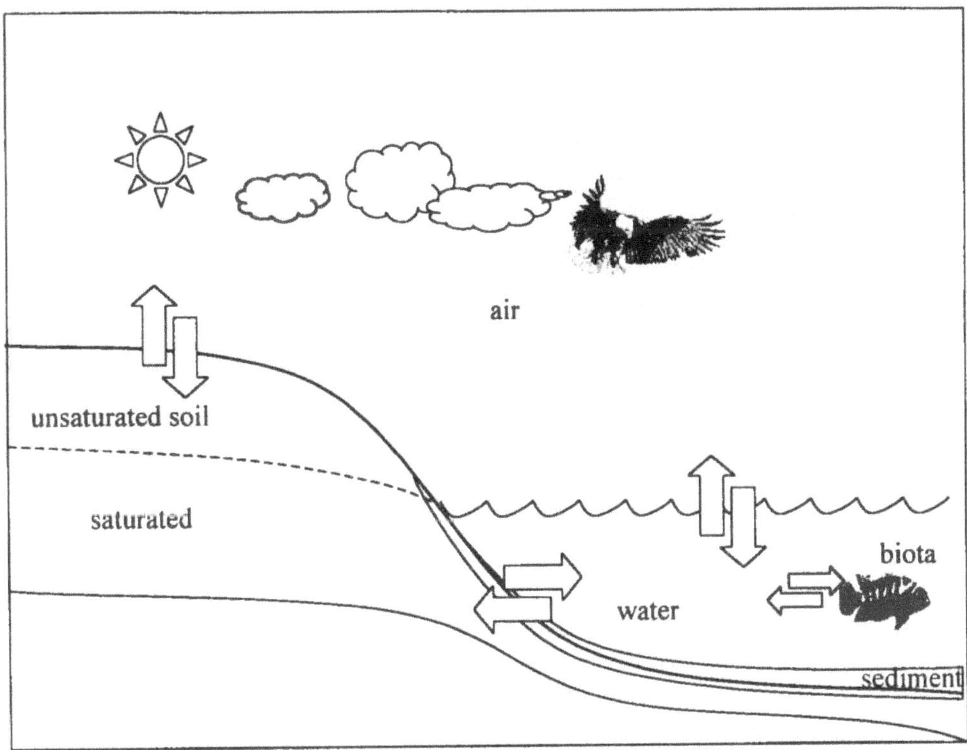

Figure 2-1 Thermodynamic properties determine the distribution and form of a chemical contaminant in the different natural phases of the environment.

2.1.1 Chemical equilibrium in air-water phases

The vapor pressure of a dissolved component above the solution is often estimated using *Henry's law*. Mathematically, Henry's law is expressed as

$$p_A = H_c x_A \tag{2-1}$$

where the ratio of the vapor pressure, p_A, to the component's concentration in the water, x_A, is given by a component-specific constant known as the Henry's law constant, H_c.

By Henry's law, the air phase concentration is assumed linearly dependent on the solute concentration in the water phase. This rule is generally obeyed only at low solute concentrations. However, this is acceptable for a wide range of volatile and semivolatile organic compounds of environmental interest.

Henry's law constants for many chemical species may be found in Perry and Green (1984), Thibodeaux (1996), or Reible (1999). Unfortunately, the constants available in the literature are reported in widely varying units. The vapor pressure is generally given as pascals or atmospheres but other measures of vapor phase concentrations including mole fraction or vapor phase concentration may be used. Liquid phase solute concentration may also be in mole fractions, mass concentrations, molar concentrations, or mass fractions. In particular, note the definition of a Henry's law constant: a ratio of mole fractions or a ratio of concentrations appearing to be dimensionless and yet having different numerical values and being used quite differently. The reader is cautioned to always check the definition of a reported Henry's law constant.

The *air-water partition coefficient* is used herein to refer to a Henry's law constant in which the contaminant loading in both the air and water phases is given as a concentration.

$$K_{air-water} = \left.\frac{c_A^{air}}{c_A^{water}}\right|_{equilibrium} \tag{2-2}$$

The concentrations are defined on either a mass or molar basis. Conversion of a vapor pressure to a mass concentration can be accomplished under environmental conditions using the ideal gas law

$$c_A^{air} = \frac{p_A \cdot MW}{R \cdot T} \tag{2-3}$$

2.1.2 Chemical equilibrium in water-organic liquid phases

Low concentration distribution between water and an organic liquid phase is usually approximated by another linear partitioning relationship. The *octanol-water partition coefficient* is an example of such a relationship and is often used to characterize the tendency to partition between water and organic phases.

$$K_{octanol-water} = \left.\frac{c_A^{octanol}}{c_A^{water}}\right|_{equilibrium} \tag{2-4}$$

Both octanol and water phases are concentrations on either a mass or molar basis. Values for the octanol-water partition coefficient may be found in Lyman et al. (1982).

Octanol was selected as the reference organic phase as it has a similar carbon-to-oxygen ratio as lipid material in animal fats. Thus it is believed that the octanol-water partitioning is representative of solute accumulation in the body (Clark, 1996).

2.1.3 Chemical equilibrium in the air-water-soil phases

Soils and sediments are a complex agglomeration of minerals (e.g., kaolinite and quartz) and organic matter (e.g., humus, plants, and animals in different stages of decay). In the unsaturated region, this solid matrix is intimately combined with air and water phases in the pore spaces. In the saturated region only water fills the pore spaces of the soil and partitioning between these phases is relatively simple, to a first approximation. In the unsaturated regime, chemical partitioning is dramatically influenced by moisture content. Dry soils may hold up to several hundred times the mass of a hydrophobic contaminant species than damp soils.

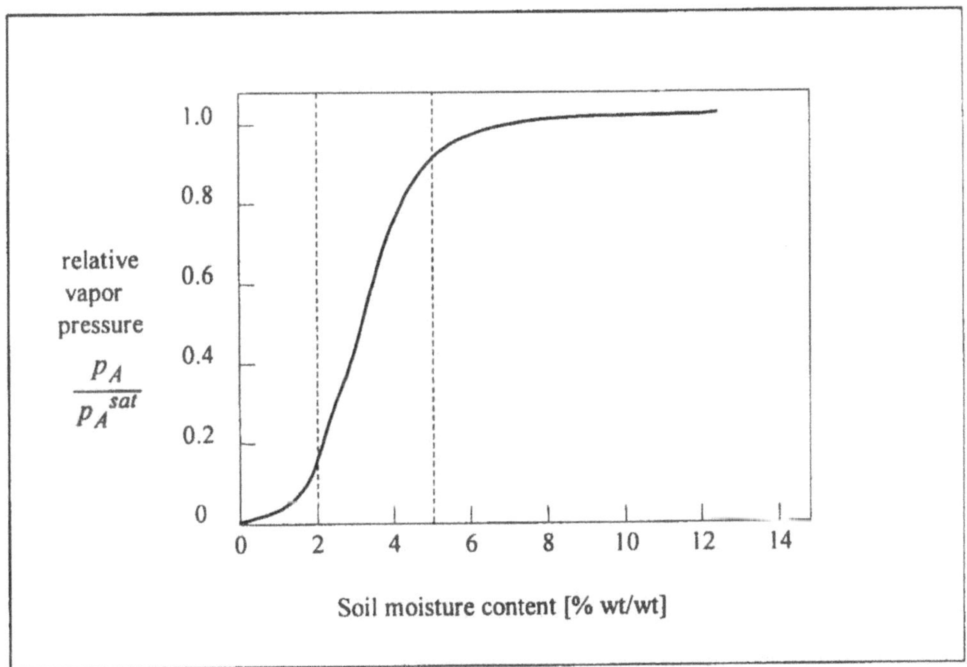

Figure 2-2 Typical equilibrium vapor pressure profile of a contaminated sediment at different moisture contents.

Valsaraj and Thibodeaux (1988) define three ranges of soil moisture content, termed dry, damp, and wet, in which the mechanisms of the uptake of organic compounds by the soil are different for each range. Linear partition coefficients have been defined for each of these zones as in the following table.

Table 2-1 Soil-air partition coefficients

Dry soil	$K_{soil-air,dry} = \frac{W_{mA}}{p_A^{sat}} \exp\left(\frac{\Delta\hat{H}_{cond} - \Delta\hat{H}_{ads}}{RT}\right)$
Damp soil	$K_{soil-air,damp} = \frac{W_{mA}}{p_A^{sat}} \exp\left(\frac{\Delta\hat{H}_{cond} - \Delta\hat{H}_{ads}}{RT}\right)\left(1 - \frac{W_B}{W_{mB}}\right)$
Wet soil	$K_{soil-air,wet} = f_{oc} \cdot K_{oc-water}\left(\frac{R \cdot T}{H_c \cdot MW}\right)$

where $K_{soil-air}$ is the soil-air partition coefficient.

$$K_{soil-air} = \left.\frac{c_A^{soil}}{c_A^{air}}\right|_{equilibrium} \tag{2-5}$$

2.1.3.1 Dry soils

Soils are considered dry when the moisture content is < ~2% by weight. Adsorption of the contaminant molecules occurs on the large surface area provided by both the mineral and organic matter of the soil particles. A large amount of adsorption is possible, consequently the vapor concentrations in equilibrium are very low.

The expression for $K_{soil-air,dry}$ in Table 2-1 is derived by considering contaminant adsorption-desorption followed the Brauner-Emmett-Teller (BET) model. A linearized, single component form of the BET isotherm was used to describe the adsoption, where

- W_{mA} is the monolayer capacity of the soil for the contaminant [g/kg(soil)]
- $\Delta\hat{H}_{abs}$ molar heat of absorption of the monolayer [J/mol]
- $\Delta\hat{H}_{cond}$ molar heat of condensation of the monolayer [J/mol]

Each of these parameters are functions of contaminant species and soil type. Values and methods of estimating these parameters are given by Valsaraj and Thibodeaux (1992).

2.1.3.2 Damp soils

Soils are considered damp when the moisture content is ~2-5% by weight. There now exists a substantial amount of water molecules (up to a monolayer coverage) to competitively adsorb onto the soil. The mineral fraction of the soil (which provides the predominant number of adsorption sites) adsorbs the water molecules in preference to the contaminant molecules due to their stronger affinity to polar molecules. This in effect causes displacement of the contaminant molecules from the soil and hence higher equilibrium vapor concentrations.

The expression for $K_{soil-air,damp}$ in Table 2-1 is derived by considering a simplified form of the two-component BET isotherm. The terms required for its calculation are the same as those in the dry case, with the addition of two terms for the competing water molecules.

- W_{mB} is the monolayer capacity of the soil for the water molecules [g/kg(soil)].
- W_B is the weight of water per unit mass of soil [g/kg(soil)].

The value of W_{mB} is estimated using the same methods as that of estimating the contaminant monolayer capacity.

2.1.3.3 *Wet soils*

Soils are considered wet when the moisture content > ~5% by weight. This includes all saturated soils and most unsaturated soils below the upper few centimeters of soil. At these moisture levels the whole soil surface area, provided by the minerals and organic matter, is covered by the polar water molecules. Adsorption of hydrophobic organics is now dominated by the organic matter component of the soil.

Thus $K_{soil-air,wet}$ is proportional to the fraction of organic carbon, f_{oc}, in the soil and the organic carbon partition coefficient, $K_{oc-water}$. f_{oc} is typically in the range of 0.02-0.04 for near-surface soils, while the organic carbon-based partition coefficient is a function of the hydrophobicity of the partitioning compound. The organic carbon partition coefficient, $K_{oc-water}$, has been given as various functions of the octanol-water partition coefficient. For example, Curtis et al. (1986) correlate the organic carbon partition as

$$\log K_{oc-water} = 0.92 \log K_{octanol-water} - 0.23 \tag{2-6}$$

Further details may be found in Valsaraj and Thibodeaux (1992).

The basis for the equilibrium partitioning relationship between soil and air in a wet soil in Table 2-1 is the soil-water partition coefficient.

$$K_{soil-water} = \left.\frac{C_A^{soil}}{C_A^{water}}\right|_{equilibrium} \approx f_{oc} K_{oc-water} \tag{2-7}$$

Note that the soil-air partition coefficient is simply the soil-water partition coefficient divided by the air-water partition coefficient.

2.2 Diffusion and the diffusion coefficient

2.2.1 Diffusion in free phases

Diffusion is an important mass transport process, especially in vertical transport in the subsurface where low permeability and pressure gradients limit advective transport. Diffusion is the term given to the mixing process due to the natural random motion of fluid particles. Mass transport by diffusion is described by *Fick's law*. For a one-dimensional system, this is mathematically expressed as

$$j_A = -\mathscr{D}_A \frac{\partial c_A}{\partial x} \qquad (2\text{-}8)$$

where the mass flux is proportional to the concentration gradient of the chemical species. The constant of proportionality is called the *diffusion coefficient*. The negative sign simply indicates that the direction of contaminant movement is in the direction of decreasing concentration (i.e., the flux is positive when the concentration gradient is negative).

The phases in which diffusion is of most interest for the environment is that of air and water. The diffusion coefficient of many chemical species in the air phase at atmospheric pressure is approximately 10^{-5} m^2s^{-1}, while diffusion coefficients in water at 20°C is 10^{-9} m^2s^{-1}.

Diffusion coefficient values for a large number of chemical species in air and water are available in Lyman et al. (1982), Sinnott (1993), and Thibodeaux (1996). Corrections for different temperatures and pressures are discussed by Bird et al. (1960). Diffusion coefficients may also be estimated using correlations of Fuller-Schttler-Giddings, for diffusion in vapor, and Wilke-Chang, for diffusion in liquids. These two methods are outlined in Appendices D and E, respectively.

2.2.2 Effective diffusion coefficient in a porous medium

An effective diffusion coefficient in a porous medium is function of the free air or free water diffusion coefficients and the physical properties of the solid matrix. Mathematically, it is defined as

$$D_{A(eff)} = \mathscr{D}_A \frac{\varepsilon}{\tau} \qquad (2\text{-}9)$$

where ε is the porosity of the medium and τ is the factor representing the tortuosity of the flow path. The tortuosity of an unconsolidated soil matrix is typically 1.2-1.5 but can be much higher.

For unsaturated soils, when the pore space contains both air and water, the effective diffusion may also be approximated by

$$D_{A(eff)} = \mathscr{D}_A \frac{\varepsilon_{air}^{10/3}}{(\varepsilon_{air} + \varepsilon_{water})^2} \qquad (2\text{-}10)$$

where ε_{air} and ε_{water} are the air-filled porosity and water-filled porosity, respectively.

Thus, for a medium containing a single mobile phase (e.g., completely dry or water-saturated soils) the effective diffusivity may be estimated by the relationship derived by Millington and Quirk (1961):

$$D_{A(eff)} = \mathscr{D}_A \, \varepsilon^{4/3} \qquad (2\text{-}11)$$

2.3 Advection and the surface mass transfer coefficient

Under conditions where it is not possible to neglect the advective component of flux, this flux can be easily defined by

$$j_A = v(C_1 - C_{A\infty}) \qquad (2\text{-}12)$$

That is, the advective flux is simply the product of the velocity, which represents a volumetric flux (volume per area per time) times the concentration of A in that volume. Advection within the transport domain is often of importance, but as indicated previously our primary focus is on diffusion in multilayered systems.

Advection also enters at the boundary of many diffusion problems. Diffusion may dominate within the domain, for example, within the soil or sediment, while at the soil or sediment surface, advection in the free fluid cannot be neglected. The complicated nature of the flux at an interface such as this is usually characterized through use of a mass transfer coefficient, an empirical coefficient that relates the concentration difference driving mass transport to the observed flux.

$$j_A = k_a\left(c_{A,surface} - c_A^{\infty}\right) \qquad (2\text{-}13)$$

The constant of proportionality is the mass transfer coefficient, k_a. The following table summarizes several correlations that may be used for predicting the surface mass transfer coefficient above sediment beds. These are limited to idealized situations but may prove useful at particular environmental interfaces as long as density variations in the fluid do not hinder or enhance buoyancy effects. More general correlations may be found in Reible (1999) or Thibodeaux (1996).

Table 2-2 Surface mass transfer coefficient correlations

Laminar flow boundary layer theory	$k_a = 0.664\,\mathrm{Re}^{0.5}\mathrm{Sc}^{1/3}\left(\frac{\mathscr{D}_A}{L_x}\right)$
Turbulent flow mass transfer theory	$k_a = 0.036\,\mathrm{Re}^{0.8}\mathrm{Sc}^{1/3}\left(\frac{\mathscr{D}_A}{L_x}\right)$
Penetration theory	$k_a = \sqrt{\frac{\mathscr{D}_A \cdot u_x}{4\pi L_x}}$

2.3.1 Laminar flow boundary layer theory and turbulent flow mass transfer

The laminar equation is an exact solution of the momentum and mass flux equations developed from a model of an isothermal (constant density) fluid flowing next to a solid wall. This model should be used only in the laminar flow regime with Re (based upon distance from the upstream edge) < 10^5. The turbulent model was also developed for constant density fluid for higher flows as measured by the Reynold's number. The velocity in these correlations is the horizontal velocity measured in the fluid above the soil or sediment interface.

2.3.2 Penetration theory

Another turbulent flow correlation, this was developed on a conceptual model in which turbulent eddies of the bulk phase concentration penetrate to the interface for a characteristic exposure time (whereby mass transfer by simple diffusion occurs) and returns. Mathematically this was described using a semi-infinite diffusion model with the exposure time approximated by u_x/L_x. This has formed the basis for models of mass transfer across a variety of environmental interfaces.

2.4 Mass balance and transport equations

Below is a diagram of a chemical spill into the sub-surface environment. Surrounding it are just a number of the possible transport pathways and reactions that may occur.

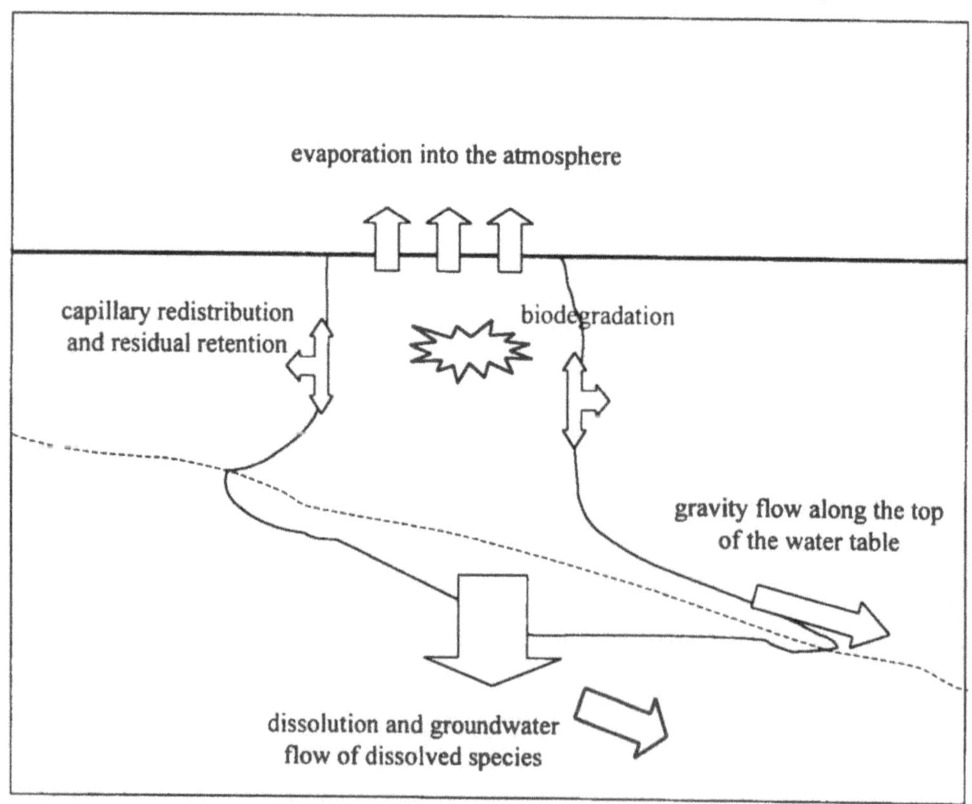

Figure 2-3 Several of the fate and transport processes acting on a chemical spill into the subsurface environment.

First we need to identify the fate and transport processes occurring that are of significance. For the components with high vapor pressures, the time in which the contaminant volatilizes into the atmosphere may be so short that we do not need to concern ourselves with biodegradation reactions. For many hydrophobic aromatic compounds in dry soil, absorption onto the soil may be so high that desorption from the soil particles and subsequent pore air/water diffusion may be the mechanisms to consider. High water solubility contaminants will lead to an advective problem flowing along with the groundwater.

Once the appropriate fate and transport mechanisms are chosen the equations for the contaminant's flux and reactions are combined in a mass balance equation

$$\left\{\begin{array}{c}\text{rate of accumulation}\\ \text{of contaminant per}\\ \text{unit volume}\end{array}\right\} = \left\{\begin{array}{c}\text{rate of}\\ \text{contaminant}\\ \text{flow in}\end{array}\right\} - \left\{\begin{array}{c}\text{rate of}\\ \text{contaminant}\\ \text{flow out}\end{array}\right\} \pm \left\{\begin{array}{c}\text{rate of}\\ \text{contaminant}\\ \text{generation / consumption}\end{array}\right\} \tag{2-14}$$

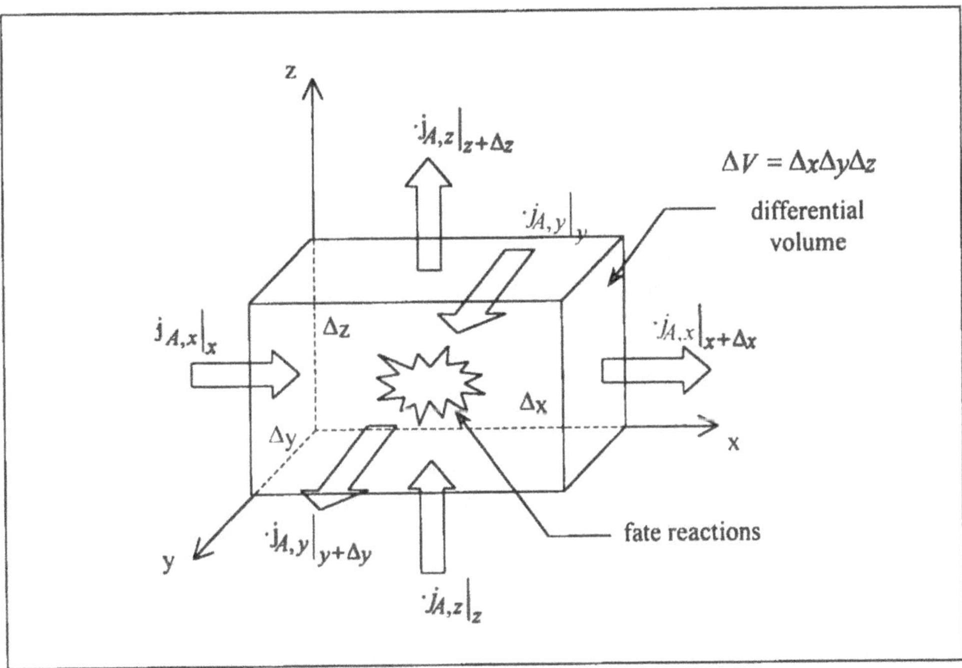

Figure 2-4 Contaminant flux through and reactions in a differential volume element.

Over a differential volume, the mass balance equation becomes

$$\frac{\partial c_A^{total}}{\partial t}\Delta x\Delta y\Delta z = \left\{ j_{A,x}\big|_x - j_{A,x}\big|_{x+\Delta x} \right\}\Delta y\Delta z + \left\{ j_{A,y}\big|_y - j_{A,y}\big|_{y+\Delta y} \right\}\Delta x\Delta z \ldots$$
$$\ldots + \left\{ j_{A,z}\big|_z - j_{A,z}\big|_{z+\Delta z} \right\}\Delta x\Delta y \pm \left(\frac{reactions}{\Delta x\Delta y\Delta z} \right) \tag{2-15}$$

Dividing the above mass balance by $\Delta x\Delta y\Delta z$ we obtain

$$\frac{\partial c_A^{total}}{\partial t} = \left\{ \frac{j_{A,x}\big|_x - j_{A,x}\big|_{x+\Delta x}}{\Delta x} \right\} + \left\{ \frac{j_{A,y}\big|_y - j_{A,y}\big|_{y+\Delta y}}{\Delta y} \right\} + \left\{ \frac{j_{A,z}\big|_z - j_{A,z}\big|_{z+\Delta z}}{\Delta z} \right\} \pm reactions \tag{2-16}$$

As we set the limits of the differential volume to zero by allowing $\Delta x \to 0, \Delta y \to 0, \Delta z \to 0$, the following differential form results:

$$\frac{\partial c_A^{total}}{\partial t} = -\nabla j_A \pm reactions \tag{2-17}$$

For a specific one-dimensional case, where transport is by diffusion only and there are no reactions the differential equation representing the contaminant's dynamics becomes

$$\frac{\partial c_A^{total}}{\partial t} = D_{A(eff)} \frac{\partial^2 c_{A,mobile\ phase}}{\partial z^2} \tag{2-18}$$

where the contaminant diffusive flux depends on the mobile phase concentration. This mobile phase is the pore air in the unsaturated zone or the pore water when modeling a system below the water table.

Let us define the *retardation factor* as

$$R_f = \frac{c_{A,total}}{c_{A,mobile\ phase}} \tag{2-19}$$

Continuing with the previous example with the diffusion occurring in an unsaturated soil system, the soil and pore water phases do not move and the contaminant's mobile phase is just the pore air. The total concentration per unit volume of soil is the sum of the contaminant masses located in each of the phases for a unit volume, that is,

$$c_A^{total} = c_A^{air}\varepsilon_{air} + c_A^{water}\varepsilon_{water} + c_A^{soil}\left(1 - \varepsilon_{air} - \varepsilon_{water}\right)\rho_{soil} \tag{2-20}$$

Assuming that the weight of the pore air makes negligible difference to the bulk (dry) density of the soil, and using our equilibrium partition relationships the total concentration may be recast as a function of the mobile phase concentration.

$$\begin{aligned} c_A^{total} &= c_A^{air}\varepsilon_{air} + c_A^{air}\frac{\varepsilon_{water}}{K_{air-water}} + c_A^{air}\rho_b K_{soil-air} \\ &= R_f c_A^{air} \end{aligned} \tag{2-21}$$

where the retardation factor for this case is given by

$$R_f = \varepsilon_{air} + \frac{\varepsilon_{water}}{K_{air-water}} + \rho_b K_{soil-air} \tag{2-22}$$

Hence, the final form of the equation that describes the dynamics of a contaminant species is stated as

$$R_f \frac{\partial c_A}{\partial t} = D_{A(eff)} \frac{\partial^2 c_A}{\partial z^2} \tag{2-23}$$

where c_A is the concentration in the air phase.

Given appropriate initial and boundary conditions, for example, mass transfer to the atmosphere at the surface and an impermeable bedrock layer a specified distance below, we can proceed to solve this equation to estimate concentrations and fluxes. Procedures for such calculations are given in Sections 3 to 7.

It must be noted, however, that the retardation factor is only defined as stated above when the parameters governing the equilibrium distribution are independent of time and space variables. A rigorous analysis of how the mass balance is modified when the air-soil equilibrium is a function of time is given in Section 9.

References

Bird, R.B., Stewart, W.E., Lightfoot, E.N. (1960) *Transport Phenomena*, Wiley, New York.

Clark, M.M. (1996) *Transport Modeling for Environmental Engineers and Scientists*, John Wiley & Sons, New York.

Curtis G.P., Reinhard, M., Roberts, P.V. (1986) *ACS Symposium Series*, **323**, 191.

Lyman, W.J., Reehl, W.F., Rosenblatt, D.H. (1982) *Handbook of Chemical Property Estimation Methods*, American Chemical Society, Washington, D.C.

Millington, R.J., Quirk, J.M. (1961) Gas Diffusion, *Trans. Faraday Society*, 57, 1200.

Schnoor, J.L. (1996) *Environmental Modeling*, John Wiley & Sons, New York.

Perry, R.H., Green, D. (1984) *Perry's Chemical Engineers' Handbook*, McGraw-Hill, Singapore.

Reible, D.D. (1999) *Fundamentals of Environmental Engineering*, CRC/Lewis Press, New York.

Thibodeaux, L.J. (1996) *Environmental Chemodynamics*, 2nd ed., John Wiley & Sons, New York.

Sinnott, R.K. (1993) *Chemical Engineering Volume 6: Design*, 2nd ed., Pergamon Press Ltd., Oxford, U.K.

Valsaraj, K.T. (1995) *Elements of Environmental Engineering*, CRC Press, Boca Raton, FL.

Valsaraj, K.T., Choy, B., Ravikrishna, R., Reible, D.D., Thibodeaux, L.J., Price, C.B., Brannon, J.M., Myers, T.E. (1997) Air emissions from exposed, contaminated sediments and dredged materials. 1. Experimental data in laboratory microcosms and mathematical modelling, *J. Haz. Materials*, **54**, 65-87.

Valsaraj, K.T., Thibodeaux, L.J. (1992) Equilibrium Adsorption of Chemical Vapors onto Soils: Model Predictions and Experimental Data, in *Fate of Pesticides and Chemicals in the Environment*, Schnoor, J.L., ed., John Wiley & Sons, New York.

Valsaraj, K.T., Thibodeaux, L.J. (1988) Equilibrium adsorption of chemical vapors on surface soils, landfills and landfarms – a review, *J. Haz. Materials*, **19**, 101.

3 Diffusion in a semi-infinite system

3.1 Introduction

Although all environmental systems have finite boundaries, many systems have dynamics that are slow enough such that they are not affected by their lower boundaries. In such cases, it may be appropriate to model the system as though it had a semi-infinite domain. In general, the numerical evaluation of an unbounded system is much simpler compared to those in which the lower boundary condition plays a role.

The solutions giving the mobile phase concentration profile and surface flux rate for several semi-infinite systems are presented. The fate and transport mechanisms of diffusion, adsorption, and first-order reactive decay are considered. These solutions are summarized in Section 3.2. A brief discussion on the numerical evaluation of these solutions is given in Section 3.3. The Laplace transform development of the solutions is outlined in Section 3.4.

3.2 Analysis summary

The transport of a contaminant species A is considered under different conditions. Sections 3.2.1 to 3.2.3 describe various cases where chemical dynamics are due to diffusion and adsorption processes. Sections 3.2.4 to 3.2.6 describe cases that include the additional fate process of decay, which is modeled using simple first-order kinetics. In all cases, $c_A(z,t)$ is the concentration of the mobile pore air or pore water phase.

3.2.1 Case 1: Semi-infinite region with uniform initial concentration and zero concentration at the surface

A system is defined by the following dynamics and boundary conditions

$$\frac{\partial c_A}{\partial t} = \left(\frac{D_{A(eff)}}{R_f}\right)\frac{\partial^2 c_A}{\partial z^2} \qquad z \in [0,\infty) \qquad (3\text{-}1)$$

$$c_A(z,t)\big|_{z=0} = 0 \qquad t > 0 \qquad (3\text{-}2)$$

$$c_A(z,t)\big|_{z\to\infty} = c_{A0} \qquad t > 0 \qquad (3\text{-}3)$$

$$c_A(z,t)\big|_{t=0} = c_{A0} \qquad z \in [0,\infty) \qquad (3\text{-}4)$$

This system is illustrated by the following diagram:

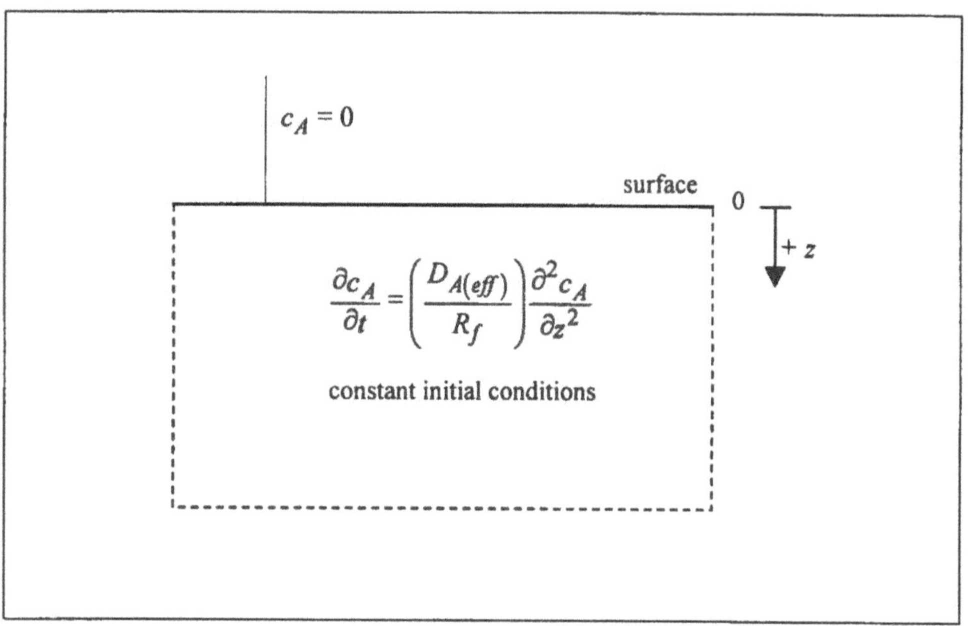

Figure 3-1 Diffusion in a semi-infinite region with uniform initial concentration and zero concentration at the surface.

The concentration profile is given by

$$c_A(z,t) = c_{A0} \cdot \operatorname{erf}\left\{\frac{z}{\sqrt{4\left(\frac{D_{A(eff)}}{R_f}\right)t}}\right\} \qquad z \in [0,\infty)\,,\, t > 0 \qquad (3\text{-}5)$$

The surface flux out of the system is given by

$$j_A(t)\big|_{z=0} = c_{A0}\sqrt{\frac{D_{A(eff)}R_f}{\pi t}} \qquad t > 0 \qquad (3\text{-}6)$$

3.2.2 Case 2: Semi-infinite region with uniform initial concentration and mass transfer or reaction at the surface

A system is defined by the following dynamics and boundary conditions:

$$\frac{\partial c_A}{\partial t} = \left(\frac{D_{A(eff)}}{R_f}\right)\frac{\partial^2 c_A}{\partial z^2} \qquad z \in [0,\infty) \qquad (3\text{-}7)$$

$$-D_{A(eff)}\frac{\partial c_A}{\partial z}\bigg|_{z=0} + k_a \cdot c_A(z,t)\big|_{z=0} = 0 \qquad t > 0 \qquad (3\text{-}8)$$

$$c_A(z,t)\big|_{z\to\infty} = c_{A0} \qquad t > 0 \qquad (3\text{-}9)$$

$$c_A(z,t)\big|_{t=0} = c_{A0} \qquad z \in [0,\infty) \qquad (3\text{-}10)$$

This system is illustrated by the following diagram:

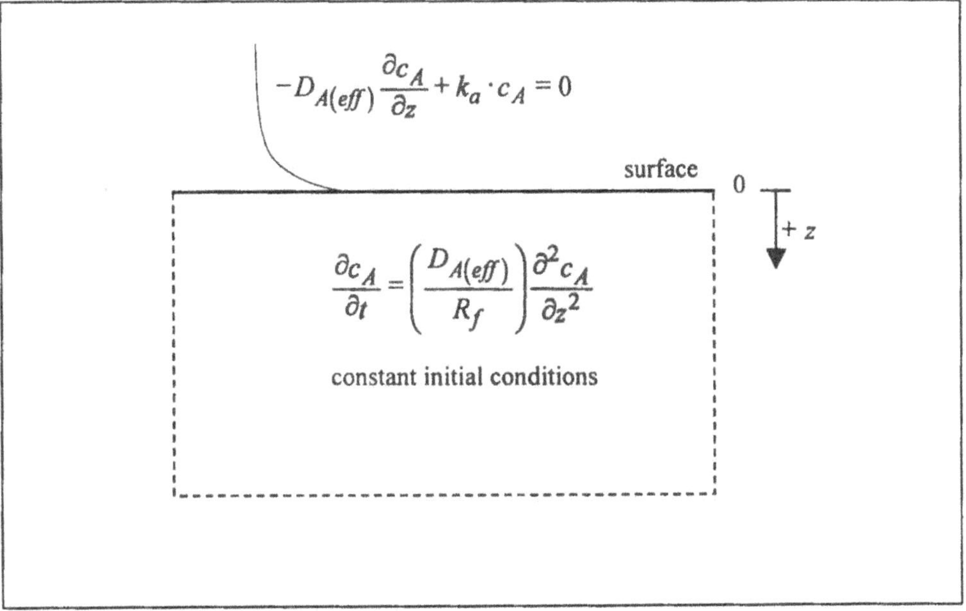

Figure 3-2 Diffusion in a semi-infinite region with uniform initial concentration and mass transfer or reaction at the surface.

The concentration profile is given by

$$c_A(z,t) = c_{A0}\left\{\begin{array}{l} \mathrm{erf}\left[\dfrac{R_f z}{\sqrt{4D_{A(eff)}R_f t}}\right] + \ldots \\ \ldots \ \exp\left(\dfrac{k_a z}{D_{A(eff)}} + \dfrac{k_a{}^2 t}{D_{A(eff)}R_f}\right)\mathrm{erfc}\left[\dfrac{R_f z}{\sqrt{4D_{A(eff)}R_f t}} + k_a\sqrt{\dfrac{t}{D_{A(eff)}R_f}}\right] \end{array}\right\}$$

$$z \in [0,\infty),\ t > 0 \qquad (3\text{-}11)$$

The surface flux out of the system is given by

$$j_A(t)\big|_{z=0} = k_a \cdot c_{A0} \cdot \exp\left(\dfrac{k_a{}^2 t}{D_{A(eff)}R_f}\right)\mathrm{erfc}\left[k_a\sqrt{\dfrac{t}{D_{A(eff)}R_f}}\right]$$

$$t > 0 \qquad (3\text{-}12)$$

3.2.3 Case 3: Semi-infinite region with uniform initial concentration capped by a finite layer with a different uniform initial concentration, and zero concentration at the surface

A system is defined by the following dynamics and boundary conditions:

$$\frac{\partial c_A}{\partial t} = \left(\frac{D_{A(eff)}}{R_f}\right)\frac{\partial^2 c_A}{\partial z^2} \qquad z \in [0,\infty) \qquad (3\text{-}13)$$

$$c_A(z,t)\big|_{z=0} = 0 \qquad t > 0 \qquad (3\text{-}14)$$

$$c_A(z,t)\big|_{z\to\infty} = c_2 \qquad t > 0 \qquad (3\text{-}15)$$

$$c_A(z,t)\big|_{t=0} = \begin{cases} c_1 & z \in [0,z_1) \\ c_2 & z \in [z_1,\infty) \end{cases} \qquad (3\text{-}16)$$

This system is illustrated by the following diagram:

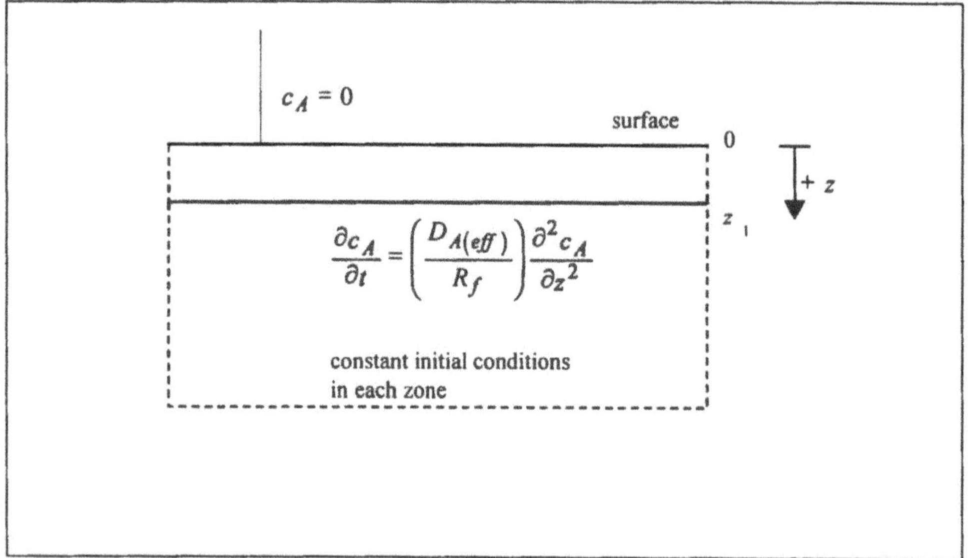

Figure 3-3 Diffusion in a semi-infinite region with uniform initial concentration capped by a finite layer with a different uniform initial concentration, and zero concentration at the surface.

The concentration profile is given by

$$c_A(z,t) = c_2 + \left(\frac{c_1 - c_2}{2}\right)\left\{\operatorname{erfc}\left[\frac{R_f(z-z_1)}{\sqrt{4D_{A(eff)}R_f t}}\right] + \operatorname{erfc}\left[\frac{R_f(z+z_1)}{\sqrt{4D_{A(eff)}R_f t}}\right]\right\} - c_1\operatorname{erfc}\left[\frac{R_f z}{\sqrt{4D_{A(eff)}R_f t}}\right]$$

$$z \in [0,\infty),\ t > 0 \qquad (3\text{-}17)$$

The surface flux out of the system is given by

$$j_A(t)\Big|_{z=0} = \sqrt{\frac{D_{A(eff)}R_f}{\pi t}}\left[\left(\frac{c_2 - c_1}{2}\right)\left\{\exp\left[-\frac{R_f z_1^{\,2}}{4D_{A(eff)}t}\right] + \exp\left[\frac{R_f z_1^{\,2}}{4D_{A(eff)}t}\right]\right\} + c_1\right] \qquad t>0 \qquad (3\text{-}18)$$

3.2.4 Case 4: Semi-infinite region with uniform initial concentration, zero concentration at the surface, and first-order decay

A system is defined by the following dynamics and boundary conditions:

$$\frac{\partial c_A}{\partial t} = \left(\frac{D_{A(eff)}}{R_f}\right)\frac{\partial^2 c_A}{\partial z^2} - k_1 c_A(z,t) \qquad z \in [0,\infty) \qquad (3\text{-}19)$$

$$c_A(z,t)\Big|_{z=0} = 0 \qquad t>0 \qquad (3\text{-}20)$$

$$c_A(z,t)\Big|_{z\to\infty} = c_{A0}\exp(-k_1 t) \qquad t>0 \qquad (3\text{-}21)$$

$$c_A(z,t)\Big|_{t=0} = c_{A0} \qquad z \in [0,\infty) \qquad (3\text{-}22)$$

The concentration profile is given by

$$c_A(z,t) = c_{A0}\cdot \operatorname{erf}\left\{\frac{z}{\sqrt{4\left(\frac{D_{A(eff)}}{R_f}\right)t}}\right\}\cdot\exp(-k_1 t) \qquad z \in [0,\infty),\ t>0 \qquad (3\text{-}23)$$

The surface flux out of the system is given by

$$j_A(t)\Big|_{z=0} = c_{A0}\exp(-k_1 t)\sqrt{\frac{D_{A(eff)}R_f}{\pi t}} \qquad t>0 \qquad (3\text{-}24)$$

This system is illustrated by the following diagram:

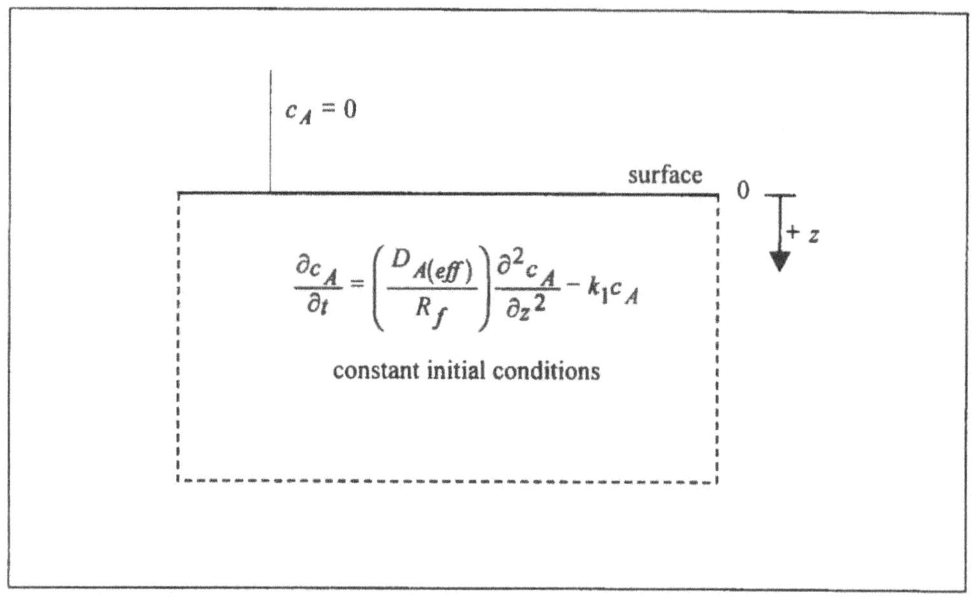

Figure 3-4 Diffusion in a semi-infinite region with uniform initial concentration, zero concentration at the surface and first-order decay.

3.2.5 Case 5: Semi-infinite region with uniform initial concentration, mass transfer or reaction at the surface, and first-order decay

A system is defined by the following dynamics and boundary conditions:

$$\frac{\partial c_A}{\partial t} = \left(\frac{D_{A(eff)}}{R_f}\right)\frac{\partial^2 c_A}{\partial z^2} - k_1 c_A(z,t) \qquad z \in [0,\infty) \tag{3-25}$$

$$-D_{A(eff)}\left.\frac{\partial c_A}{\partial z}\right|_{z=0} + k_a \cdot c_A(z,t)\big|_{z=0} = 0 \qquad t > 0 \tag{3-26}$$

$$c_A(z,t)\big|_{z\to\infty} = c_{A0}\exp(-k_1 t) \qquad t > 0 \tag{3-27}$$

$$c_A(z,t)\big|_{t=0} = c_{A0} \qquad z \in [0,\infty) \tag{3-28}$$

The concentration profile is given by

$$c_A(z,t) = c_{A0}\left\{\begin{array}{l} \operatorname{erf}\left[\dfrac{R_f z}{\sqrt{4D_{A(eff)}R_f t}}\right] + \ldots \\ \ldots \exp\left(\dfrac{k_a z}{D_{A(eff)}} + \dfrac{k_a^2 t}{D_{A(eff)}R_f}\right)\operatorname{erfc}\left[\dfrac{R_f z}{\sqrt{4D_{A(eff)}R_f t}} + k_a\sqrt{\dfrac{t}{D_{A(eff)}R_f}}\right] \end{array}\right\}\exp(-k_1 t)$$

$$z \in [0,\infty),\ t > 0 \tag{3-29}$$

The surface flux out of the system is given by

$$j_A(t)\big|_{z=0} = k_a \cdot c_{A0} \cdot \exp\left(\frac{k_a^{\,2} t}{D_{A(eff)} R_f} - k_1 t\right) \operatorname{erfc}\left[k_a \sqrt{\frac{t}{D_{A(eff)} R_f}}\right] \qquad t>0 \qquad (3\text{-}30)$$

This system is illustrated by the following diagram:

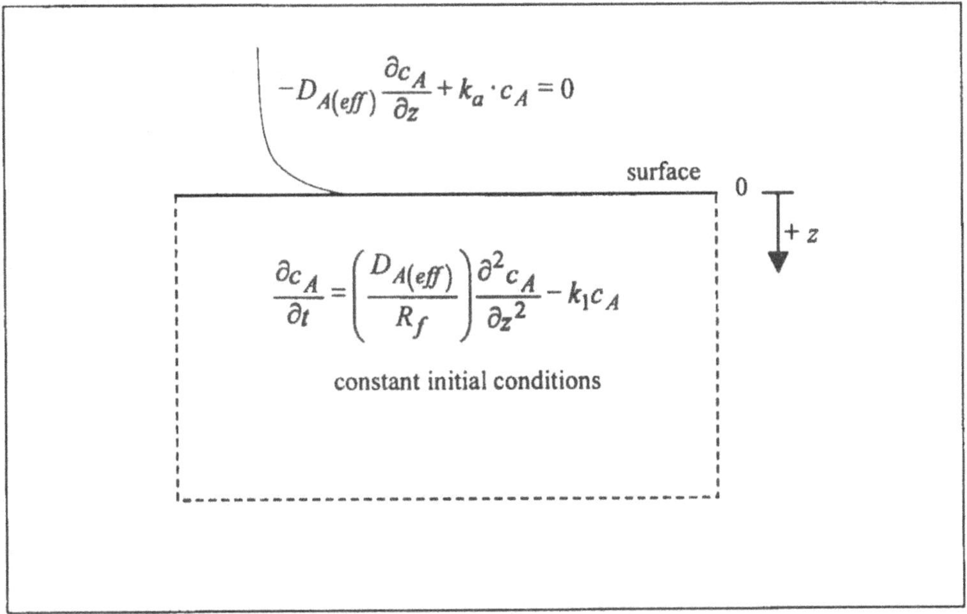

Figure 3-5 Diffusion in a semi-infinite region with uniform initial concentration, mass transfer or reaction at the surface and first-order decay.

3.2.6 Case 6: Semi-infinite region with uniform initial concentration capped by a finite layer with a different uniform initial concentration, zero concentration at the surface, and first-order decay

A system is defined by the following dynamics and boundary conditions:

$$\frac{\partial c_A}{\partial t} = \left(\frac{D_{A(eff)}}{R_f}\right)\frac{\partial^2 c_A}{\partial z^2} - k_1 c_A(z,t) \qquad z \in [0,\infty) \qquad (3\text{-}31)$$

$$c_A(z,t)\big|_{z=0} = 0 \qquad t>0 \qquad (3\text{-}32)$$

$$c_A(z,t)\big|_{z\to\infty} = c_2 \exp(-k_1 t) \qquad t>0 \qquad (3\text{-}33)$$

$$c_A(z,t)\big|_{t=0} = \begin{cases} c_1 & z \in [0, z_1) \\ c_2 & z \in [z_1, \infty) \end{cases} \qquad (3\text{-}34)$$

This system is illustrated by the following diagram:

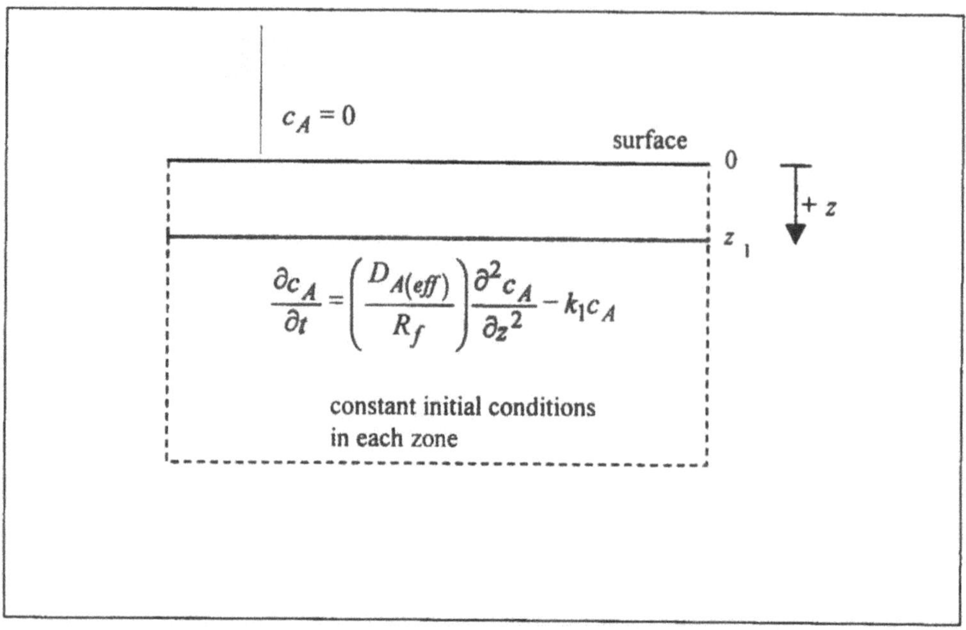

Figure 3-6 Diffusion in a semi-infinite region with uniform initial concentration capped by a finite layer with a different uniform initial concentration, zero concentration at the surface, and first-order decay.

The concentration profile is given by

$$c_A(z,t) = \left\{ \begin{array}{l} c_2 - c_1 \,\mathbf{erfc}\left[\dfrac{R_f z}{\sqrt{4D_{A(eff)} R_f t}}\right] \\ + \left(\dfrac{c_1 - c_2}{2}\right)\left\{ \mathbf{erfc}\left[\dfrac{R_f (z - z_1)}{\sqrt{4D_{A(eff)} R_f t}}\right] + \mathbf{erfc}\left[\dfrac{R_f (z + z_1)}{\sqrt{4D_{A(eff)} R_f t}}\right]\right\} \end{array} \right\} \mathbf{exp}(-k_1 t)$$

$$z \in [0, \infty), t > 0 \qquad \text{(3-35)}$$

The surface flux out of the system is given by

$$j_A(t)\big|_{z=0} = \sqrt{\frac{D_{A(eff)} R_f}{\pi t}} \left[\left(\frac{c_2 - c_1}{2}\right) \left\{ \mathbf{exp}\left[-\frac{R_f z_1^2}{4D_{A(eff)} t}\right] + \mathbf{exp}\left[\frac{R_f z_1^2}{4D_{A(eff)} t}\right]\right\} + c_1 \right] \mathbf{exp}(-k_1 t)$$

$$t > 0 \qquad \text{(3-36)}$$

3.3 Numerical Evaluation

Numerical evaluation of the concentration profile and some of the surface flux expressions requires the calculation of the error function or the complimentary error function. The *error function* is defined as

$$\operatorname{erf}(u) = \frac{2}{\sqrt{\pi}} \int_0^u e^{-\eta^2} d\eta \tag{3-37}$$

and the *complimentary error function* as

$$\operatorname{erfc}(u) = 1 - \operatorname{erf}(u) = \frac{2}{\sqrt{\pi}} \int_u^\infty e^{-\eta^2} d\eta \tag{3-38}$$

Tabulations of the error function may be found in advanced calculus textbooks and other mathematics handbooks. Error function values can also be obtained from the other sources including Microsoft Excel (with the analysis tool pack add-in), Mathcad, IMSL Fortran Math Library Special Functions (call ERF(X)), Matlab, and other mathematical software packages. Further details about the error function can be found in Appendix A.

Numerical evaluation of complementary error function, $\operatorname{erfc}(u)$, at high values of u may cause problems in cases 2 and 5 as the numerical limit is reached. As u increases, the complementary error function tends toward zero. However, an absolute zero value in the flux equations produces erroneous results. High values of u occur at high values of k_a, t, and z. In all these situations, the solution tends toward the equivalent zero concentration boundary condition cases of 1 and 4, respectively. These solutions should be used when numerical instability is encountered.

3.4 Development

3.4.1 Laplace transformation method

Partial differential equation in terms of space, z, and time, t, is transformed into the Laplace domain. This results in a complex ordinary differential equation in the space domain only. This equation in now solved and transformed back into the time domain. Further details about Laplace transformations are given in Appendix B.

Given a system defined in case 1, with the partial differential equation (3-1) and initial conditions given by (3-4), the Laplace transformation results in

$$\mathscr{L}\left[\frac{\partial c_A}{\partial t}\right] = \mathscr{L}\left[\left(\frac{D_{A(eff)}}{R_f}\right)\frac{\partial^2 c_A}{\partial z^2}\right] \tag{3-39}$$

$$\therefore \quad s \cdot C(z,s) - c_{A0} = \left(\frac{D_{A(eff)}}{R_f}\right)\frac{d^2 C}{dz^2} \tag{3-40}$$

where $C(z,s)$ is the Laplace transform of $c_A(z,t)$.

The boundary condition (3-2) is transformed to give

$$\mathscr{L}\left[c_A(z,t)\right] = \mathscr{L}[0] \qquad z = 0 \qquad (3\text{-}41)$$

$$\therefore \quad C(z,s) = 0 \qquad z = 0 \qquad (3\text{-}42)$$

The general solution for equation (3-40) has the form

$$C(z,s) = A \cdot \exp\left(-z\sqrt{\frac{R_f s}{D_{A(eff)}}}\right) + B \cdot \exp\left(z\sqrt{\frac{R_f s}{D_{A(eff)}}}\right) + \frac{c_{A0}}{s} \qquad (3\text{-}43)$$

As $C(z,s)$ is required to be bounded for a feasible solution in the time domain, $B = 0$. Using the boundary condition (3-42), we solve for coefficient A:

$$A = -\frac{c_{A0}}{s} \qquad (3\text{-}44)$$

Hence the solution to the system in the Laplace domain is given by

$$C(z,s) = c_{A0}\left[\frac{1}{s} - \frac{1}{s}\exp\left(-\sqrt{\frac{R_f}{D_{A(eff)}}}z \cdot \sqrt{s}\right)\right] \qquad (3\text{-}45)$$

Using the tables of Laplace transforms in Appendix B, the equivalent solution in the time domain is given by

$$\begin{aligned} c_A(z,t) &= c_{A0}\left[1 - \operatorname{erfc}\left(z\sqrt{\frac{R_f}{4D_{A(eff)}t}}\right)\right] \\ &= c_{A0}\operatorname{erf}\left(z\sqrt{\frac{R_f}{4D_{A(eff)}t}}\right) \end{aligned} \qquad (3\text{-}46)$$

To obtain the flux from the boundary $z = 0$ out of the system, differentiate equation (3-46) with respect to z at $z = 0$:

$$
\begin{aligned}
j_A(t)\big|_{z=0} &= D_{A(eff)}\frac{\partial c_A}{\partial z}\bigg|_{z=0} \\
&= c_{A0}\cdot D_{A(eff)}\cdot\frac{2}{\sqrt{\pi}}\left\{\frac{d}{dz}\int_0^{z/\sqrt{4D_{A(eff)}t/R_f}}\exp\left(-\eta^2\right)d\eta\right\}\Bigg|_{z=0} \\
&= c_{A0}\cdot D_{A(eff)}\cdot\frac{2}{\sqrt{\pi}}\left\{\frac{1}{\sqrt{4D_{A(eff)}t/R_f}}\exp\left(-\left(\frac{z}{\sqrt{4D_{A(eff)}t/R_f}}\right)^2\right)\right\}\Bigg|_{z=0} \\
&= \frac{c_{A0}\cdot D_{A(eff)}}{\sqrt{\pi D_{A(eff)}t/R_f}} \\
&= c_{A0}\sqrt{\frac{D_{A(eff)}R_f}{\pi t}}
\end{aligned}
\tag{3-47}
$$

Case 2 is solved in a similar fashion. The transformation of the partial differential equation (3-7) of the boundary conditions, (3-8) and (3-9), gives the following:

$$s\cdot C(z,s)-c_{A0}=\left(\frac{D_{A(eff)}}{R_f}\right)\frac{d^2C}{dz^2} \qquad z\in[0,\infty) \tag{3-48}$$

$$-D_{A(eff)}\frac{dC}{dz}+k_a\cdot C(z,s)=0 \qquad z=0 \tag{3-49}$$

$$C(z,s)=\frac{c_{A0}}{s} \qquad z\to\infty \tag{3-50}$$

The general solution for equation (3-48) has the form

$$C(z,s)=A\cdot\exp\left(-z\sqrt{\frac{R_f s}{D_{A(eff)}}}\right)+\frac{c_{A0}}{s} \tag{3-51}$$

where the coefficient of the positive exponent is once again zero as $C(z,s)$ is required to be bounded for a feasible solution in the time domain. Hence

$$\frac{dC}{dz}=-A\sqrt{\frac{R_f s}{D_{A(eff)}}}\cdot\exp\left(-z\sqrt{\frac{R_f s}{D_{A(eff)}}}\right) \tag{3-52}$$

Substituting equations (3-51) and (3-52) into the boundary condition (3-49) at $z = 0$ to solve for A,

$$-D_{A(eff)}\left[-A\sqrt{\frac{R_f s}{D_{A(eff)}}}\right]+k_a\left[A+\frac{c_{A0}}{s}\right]=0$$

$$\therefore \quad A=\frac{c_{A0}\left(\dfrac{k_a}{\sqrt{D_{A(eff)}R_f}}\right)}{s\left(\dfrac{k_a}{\sqrt{D_{A(eff)}R_f}}+\sqrt{s}\right)} \tag{3-53}$$

Thus the solution in the Laplace domain is given by

$$C(z,s)=c_{A0}\left[\frac{1}{s}-\frac{\dfrac{k_a}{\sqrt{D_{A(eff)}R_f}}}{s\left[\dfrac{k_a}{\sqrt{D_{A(eff)}R_f}}+\sqrt{s}\right]}\cdot\exp\left(-z\sqrt{\frac{R_f s}{D_{A(eff)}}}\right)\right] \tag{3-54}$$

Inverting the transformation using the table of Laplace transforms in Appendix B we obtain

$$c_A(z,t)=c_{A0}\left\{\begin{array}{l}\operatorname{erf}\left[\dfrac{R_f z}{\sqrt{4D_{A(eff)}R_f t}}\right]+\ldots \\ \ldots\ \exp\left(\dfrac{k_a z}{D_{A(eff)}}+\dfrac{k_a^2 t}{D_{A(eff)}R_f}\right)\operatorname{erfc}\left[\dfrac{R_f z}{\sqrt{4D_{A(eff)}R_f t}}+k_a\sqrt{\dfrac{t}{D_{A(eff)}R_f}}\right]\end{array}\right\}$$

(3-55)

To obtain the flux from the surface out of the system, set $z = 0$, and multiply by the mass transfer coefficient giving

$$j_A(t)\big|_{z=0} = k_a \cdot c_{A0}\left\{\begin{array}{l}\operatorname{erf}\left[\dfrac{R_f z}{\sqrt{4D_{A(eff)}R_f t}}\right]+\ldots \\ \ldots\exp\left(\dfrac{k_a z}{D_{A(eff)}}+\dfrac{k_a{}^2 t}{D_{A(eff)}R_f}\right)\operatorname{erfc}\left[\dfrac{R_f z}{\sqrt{4D_{A(eff)}R_f t}}+k_a\sqrt{\dfrac{t}{D_{A(eff)}R_f}}\right]\end{array}\right\}\Bigg|_{z=0}$$

$$= k_a \cdot c_{A0} \cdot \exp\left(\frac{k_a{}^2 t}{D_{A(eff)}R_f}\right)\operatorname{erfc}\left[k_a\sqrt{\frac{t}{D_{A(eff)}R_f}}\right] \tag{3-56}$$

3.4.2 Principle of superposition

As the diffusion equation is linear, if a problem with the same linear boundary conditions may be decomposed into simpler sub-problems, the overall solution is the sum of the solutions to the constituent sub-problems. This is known as the principle of superposition.

Case 3 may be considered as a variation of case 1 and the following auxiliary problem:

$$\frac{\partial c_A}{\partial t} = \left(\frac{D_{A(eff)}}{R_f}\right)\frac{\partial^2 c_A}{\partial z^2} \qquad z \in [0,\infty) \tag{3-57}$$

$$c_A(z,t)\big|_{z=0} = 0 \qquad t > 0 \tag{3-58}$$

$$c_A(z,t)\big|_{z\to\infty} = 0 \qquad t > 0 \tag{3-59}$$

$$c_A(z,t)\big|_{t=0} = \begin{cases} c_1 - c_2 & z \in [0, z_1) \\ 0 & z \in [z_1, \infty) \end{cases} \tag{3-60}$$

The solution to the auxiliary problem posed by equations (3-57) to (3-60) is given by Carslaw and Jaeger (1959) as

$$c_A(z,t) = \left(\frac{c_1 - c_2}{2}\right)\left[\operatorname{erfc}\left(\frac{R_f(z-z_1)}{\sqrt{4D_{A(eff)}R_f t}}\right)+\operatorname{erfc}\left(\frac{R_f(z+z_1)}{\sqrt{4D_{A(eff)}R_f t}}\right)-2\operatorname{erfc}\left(\frac{R_f z}{\sqrt{4D_{A(eff)}R_f t}}\right)\right] \tag{3-61}$$

The solution to case 1 (with initial concentration of c_2) is given by

$$c_A(z,t) = c_2\left[1-\operatorname{erfc}\left(z\sqrt{\frac{R_f}{4D_{A(eff)}t}}\right)\right] \tag{3-62}$$

Superimposing equations (3-61) and (3-62) we obtain the solution to the combines system:

$$c_A(z,t) = c_2 + \left(\frac{c_1 - c_2}{2}\right)\left\{\mathrm{erfc}\left[\frac{R_f(z-z_1)}{\sqrt{4D_{A(eff)}R_f t}}\right] + \mathrm{erfc}\left[\frac{R_f(z+z_1)}{\sqrt{4D_{A(eff)}R_f t}}\right]\right\} - c_1\,\mathrm{erfc}\left[\frac{R_f z}{\sqrt{4D_{A(eff)}R_f t}}\right]$$

(3-63)

To obtain the flux from the boundary $z = 0$ out of the system, differentiate equation (3-63) with respect to z at $z = 0$:

$$j_A(t)\Big|_{z=0} = D_{A(eff)}\frac{\partial}{\partial z}\left[\begin{array}{l}\left(\frac{c_1 - c_2}{2}\right)\left\{\mathrm{erfc}\left[\frac{R_f(z-z_1)}{\sqrt{4D_{A(eff)}R_f t}}\right] + \mathrm{erfc}\left[\frac{R_f(z+z_1)}{\sqrt{4D_{A(eff)}R_f t}}\right]\right\} \\ +c_2 - c_1\,\mathrm{erfc}\left[\frac{R_f z}{\sqrt{4D_{A(eff)}R_f t}}\right]\end{array}\right]_{z=0}$$

$$= \left[\begin{array}{l}\left(\frac{c_1 - c_2}{2}\right)\sqrt{\frac{D_{A(eff)}R_f}{\pi t}}\left\{-\exp\left[-\frac{R_f(z-z_1)^2}{4D_{A(eff)}t}\right] - \exp\left[-\frac{R_f(z+z_1)^2}{4D_{A(eff)}t}\right]\right\} \\ +c_1\sqrt{\frac{D_{A(eff)}R_f}{\pi t}}\exp\left[-\frac{R_f z^2}{4D_{A(eff)}t}\right]\end{array}\right]_{z=0}$$

$$= \sqrt{\frac{D_{A(eff)}R_f}{\pi t}}\left[\left(\frac{c_2 - c_1}{2}\right)\left\{\exp\left[-\frac{R_f z_1^2}{4D_{A(eff)}t}\right] + \exp\left[\frac{R_f z_1^2}{4D_{A(eff)}t}\right]\right\} + c_1\right]$$

(3-64)

3.4.3 Variable transformation for first-order decay

Cases 4 to 6 are similar problems to that of cases 1 to 3, respectively, except the standard diffusion-adsorption dynamics now includes an additional first-order reaction term. The partial differential equation has the form

$$\frac{\partial c_A}{\partial t} = \left(\frac{D_{A(eff)}}{R_f}\right)\frac{\partial^2 c_A}{\partial z^2} - k_1 c_A(z,t) \qquad z \in [0,\infty) \qquad (3\text{-}65)$$

To solve the above, we use the following variable substitution:

$$c_A(z,t) = \hat{c}(z,t)\exp(-k_1 t) \qquad (3\text{-}66)$$

The transformed left-hand side of equation (3-65) gives

$$\frac{\partial c_A}{\partial t} = -k_1\hat{c}(z,t)\exp(-k_1 t) + \frac{\partial \hat{c}}{\partial t}\exp(-k_1 t) \qquad (3\text{-}67)$$

The transformed right-hand side gives

$$\left(\frac{D_{A(eff)}}{R_f}\right)\frac{\partial^2 c_A}{\partial z^2}-k_1 c_A(z,t)=\left(\frac{D_{A(eff)}}{R_f}\right)\frac{\partial^2 \hat{c}}{\partial z^2}\exp(-k_1 t)-k_1\hat{c}(z,t)\exp(-k_1 t) \quad (3\text{-}68)$$

Hence, the transformed dynamic equations result in

$$\therefore \quad \frac{\partial \hat{c}}{\partial t}=\left(\frac{D_{A(eff)}}{R_f}\right)\frac{\partial^2 \hat{c}}{\partial z^2} \quad (3\text{-}69)$$

The net effect of the transformation is to remove the reaction term from the dynamic equation, resulting in the standard diffusion-adsorption formulation.

Effect of transformation on initial conditions

At time equal to zero the transformed initial concentration results in

$$c_A(z,0)=\hat{c}(z,0)\exp(0)\equiv\hat{c}(z,0) \quad (3\text{-}70)$$

Hence, all initial conditions remain unaltered after the transformation.

Effect of the transformation on boundary conditions

The general transformation of the bottom boundary results in

$$\hat{c}\exp(-k_1 t)=c_0\exp(-k_1 t) \qquad \text{at } z\to\infty,\ t>0 \quad (3\text{-}71)$$

Transformation of the upper boundary condition for the case of zero surface concentration gives

$$\hat{c}(z,t)\exp(-k_1 t)=0 \qquad \text{at } z=L,\ t>0 \quad (3\text{-}72)$$

and for a mass transfer upper boundary gives

$$D_{A(eff)}\frac{\partial \hat{c}}{\partial z}\exp(-k_1 t)+k_a\cdot\hat{c}(z,t)\exp(-k_1 t)=0 \qquad \text{at } z=L,\ t>0 \quad (3\text{-}73)$$

Dividing all these equations by the common factor $\exp(-k_1 t)$ shows that all the boundary conditions remain unaltered after the transformation, except the bottom boundary condition which now tends toward a constant value.

Effect on the solution and surface flux

As the transformed system of equations results in the standard formulation with diffusion and adsorption mechanisms only, we solve for $\hat{c}(z,t)$ as described in the previous sections of 3.4 for cases 1 to 3. The resulting solution then is inverted back into terms of $c_A(z,t)$ by equation (3-66).

The net effect of the decay term is to include an additional exponential time term to the solution. Since the flux is determined by a differential with respect to space, it too is only modified by this additional factor.

References

Özisk, M.N. (1993) *Heat Conduction*, 2nd ed., John Wiley & Sons, New York.

Carslaw, H.S., Jaeger, J.C. (1959) *Conduction of Heat in Solids*, Clarendon Press, Oxford.

Crank, J. (1975) *The Mathematics of Diffusion*, Clarendon Press, Oxford.

Farlow, S.J. (1993) *Partial Differential Equations for Scientists and Engineers*, Dover Publications Inc., New York.

4 Diffusion in a finite layer

4.1 Introduction

Many systems are often better described by finite layer geometry. Examples include experimental microcosms, and field systems that have an impermeable layer (e.g., rock). The ability to easily specify arbitrary initial concentration profiles is another important advantage of using the finite layer system models over semi-infinite domain models.

The solutions giving the mobile phase concentration profile and surface flux rate for several finite layer systems are presented. The fate and transport mechanisms of diffusion, adsorption, and first-order reactive decay are considered with a variety of surface boundary conditions. These solutions are summarized in Section 4.2. A brief discussion on the numerical evaluation of these solutions is given in Section 4.3. The system equations were solved using a separation of variables technique. The development of the solutions is outlined in Section 4.4.

4.2 Analysis summary

The dynamics of a contaminant species A due to diffusion and adsorption processes, in a finite soil or sediment layer $z \in [0, L]$, is defined by the following equation:

$$\frac{\partial c_A}{\partial t} = \left(\frac{D_{A(eff)}}{R_f} \right) \frac{\partial^2 c_A}{\partial z^2} \qquad z \in [0, L] \tag{4-1}$$

where $c_A(z,t)$ is the concentration of species A in the mobile pore air or pore water phase. Solutions for several different boundary conditions with the above partial differential equation are given in Sections 4.2.1 to 4.2.4.

If a first-order decay mechanism of the contaminant is also of significance, the governing dynamic equation has the form

$$\frac{\partial c_A}{\partial t} = \left(\frac{D_{A(eff)}}{R_f} \right) \frac{\partial^2 c_A}{\partial z^2} - k_1 c_A(z,t) \qquad z \in [0, L] \tag{4-2}$$

Solutions for several different boundary conditions with the above partial differential equation are given in Sections 4.2.5 to 4.2.8.

In both dynamic situations, the surface flux to the top interface, $z = L$, is given by

$$j_A(t)\Big|_{z=L} = -D_{A(eff)} \frac{\partial c_A}{\partial z}\Big|_{z=L} \tag{4-3}$$

4.2.1 Case 1: Finite layer with arbitrary initial concentrations, zero concentration at the surface, and zero flux at the base

A system is defined by the dynamic equation (4-1) with the following boundary and initial conditions:

$$\left.\frac{\partial c_A}{\partial z}\right|_{z=0} = 0 \qquad t > 0 \qquad (4\text{-}4)$$

$$\left.c_A(z,t)\right|_{z=L} = 0 \qquad t > 0 \qquad (4\text{-}5)$$

$$\left.c_A(z,t)\right|_{t=0} = c_{A0}(z) \qquad z \in [0, L] \qquad (4\text{-}6)$$

This system is illustrated by the following diagram:

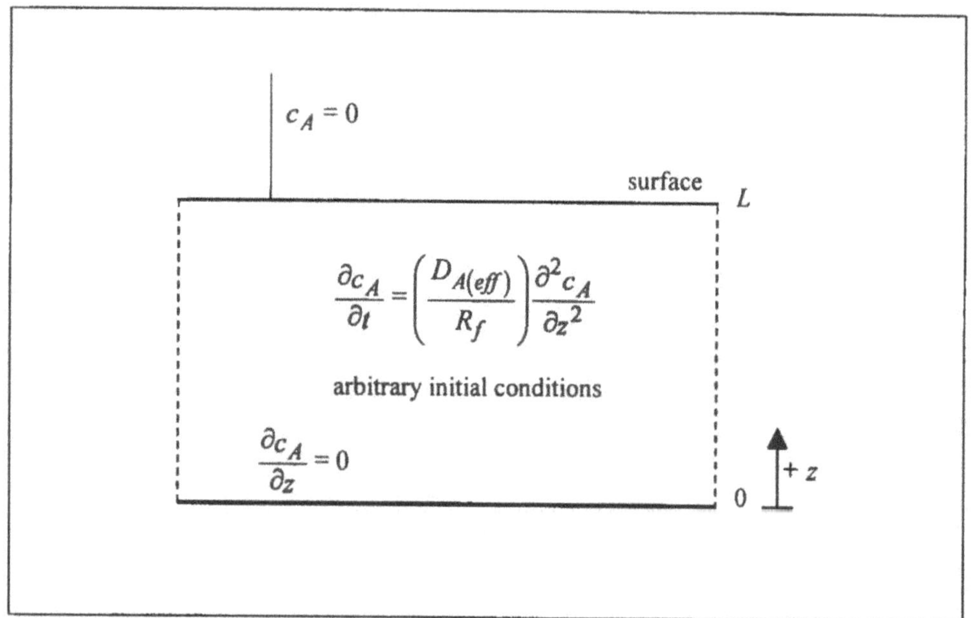

Figure 4-1 Diffusion in a finite layer with arbitrary initial concentrations, zero concentration at the surface, and zero flux at the base.

The following expression is used for calculating the concentration

$$c_A(z,t) = \frac{2}{L}\sum_{n=1}^{\infty}\left\{\exp\left(-\frac{D_{A(eff)}}{R_f}\beta_n^{\,2}t\right)\cdot\cos(\beta_n z)\cdot\int_0^L c_{A0}(z')\cdot\cos(\beta_n z')dz'\right\} \qquad (4\text{-}7)$$

and calculating the surface flux

$$j_A(t)\Big|_{z=L} = -\frac{2D_{A(eff)}}{L}\sum_{n=1}^{\infty}\left\{\exp\left(-\frac{D_{A(eff)}}{R_f}\beta_n^{\ 2}t\right)\cdot\beta_n\cdot(-1)^n\cdot\int_0^L c_{A0}(z')\cdot\cos(\beta_n z')dz'\right\} \tag{4-8}$$

where β_n is given by

$$\beta_n = \frac{\pi}{L}\left(\frac{2n-1}{2}\right) \tag{4-9}$$

4.2.2 Case 2: Finite layer with uniform initial concentration, zero surface concentration, and zero flux at the base

A system is defined by the dynamic equation (4-1) with the following boundary and initial conditions:

$$\left.\frac{\partial c_A}{\partial z}\right|_{z=0} = 0 \qquad t > 0 \tag{4-10}$$

$$c_A(z,t)\Big|_{z=L} = 0 \qquad t > 0 \tag{4-11}$$

$$c_A(z,t)\Big|_{t=0} = c_{A0} \qquad z \in [0, L] \tag{4-12}$$

This system is illustrated by the following diagram:

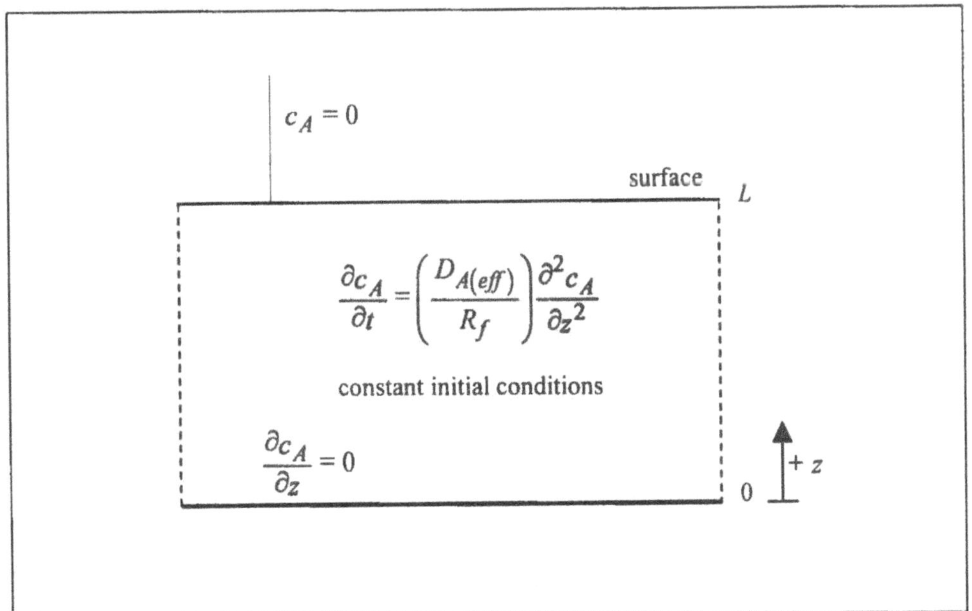

Figure 4-2 Diffusion in a finite layer with uniform initial concentration, zero surface concentration, and zero flux at the base.

The following expression is used for calculating the concentration

$$c_A(z,t) = -\frac{2c_{A0}}{L}\sum_{n=1}^{\infty}\left\{\exp\left(-\frac{D_{A(eff)}}{R_f}\beta_n{}^2 t\right)\cdot\cos(\beta_n z)\cdot\frac{1}{\beta_n}\cdot(-1)^n\right\} \qquad (4\text{-}13)$$

and calculating the surface flux

$$j_A(t)\Big|_{z=L} = \frac{2D_{A(eff)}c_{A0}}{L}\sum_{n=1}^{\infty}\exp\left(-\frac{D_{A(eff)}}{R_f}\beta_n{}^2 t\right) \qquad (4\text{-}14)$$

where β_n is given by

$$\beta_n = \frac{\pi}{L}\left(\frac{2n-1}{2}\right) \qquad (4\text{-}15)$$

4.2.3 Case 3: Finite layer with arbitrary initial concentrations, mass transfer or reaction at the surface, and zero flux at the base

A system is defined by the dynamic equation (4-1) with the following conditions:

$$\left.\frac{\partial c_A}{\partial z}\right|_{z=0} = 0 \qquad t>0 \qquad (4\text{-}16)$$

$$D_{A(eff)}\left.\frac{\partial c_A}{\partial z}\right|_{z=L} + k_a\cdot c_A(z,t)\Big|_{z=L} = 0 \qquad t>0 \qquad (4\text{-}17)$$

$$c_A(z,t)\Big|_{t=0} = c_{A0}(z) \qquad z\in[0,L] \qquad (4\text{-}18)$$

This system is illustrated by the following diagram:

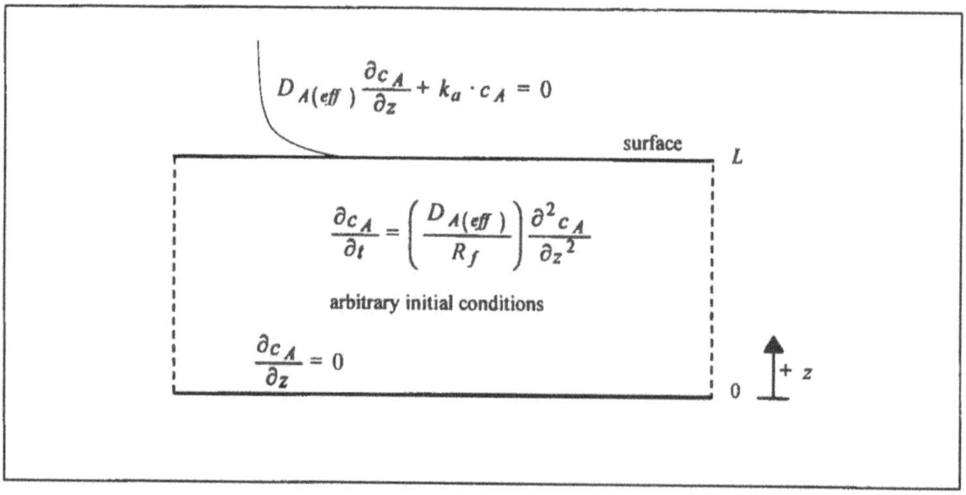

Figure 4-3 Diffusion in a finite layer with arbitrary initial concentrations, mass transfer or reaction at the surface, and zero flux at the base.

The following expression is used for calculating the concentration

$$c_A(z,t) = \frac{2}{L}\sum_{n=1}^{\infty}\left\{\exp\left(-\frac{D_{A(eff)}}{R_f}\beta_n{}^2 t\right)\cdot\cos(\beta_n z)\cdot\int_0^L c_{A0}(z')\cdot\cos(\beta_n z')dz'\right\} \quad (4\text{-}19)$$

and calculating the surface flux

$$j_A(t)\big|_{z=L} = \frac{2D_{A(eff)}}{L}\sum_{n=1}^{\infty}\left\{\exp\left(-\frac{D_{A(eff)}}{R_f}\beta_n{}^2 t\right)\cdot\beta_n\cdot\sin(\beta_n L)\cdot\int_0^L c_{A0}(z')\cdot\cos(\beta_n z')dz'\right\} \quad (4\text{-}20)$$

where β_n is given by the positive roots of

$$\beta_n\tan(\beta_n L) = \left(\frac{k_a}{D_{A(eff)}}\right) \quad (4\text{-}21)$$

4.2.4 Case 4: Finite layer with uniform initial concentration, mass transfer or reaction at the surface, and zero flux at the base

A system is defined by the dynamic equation (4-1) with the following conditions:

$$\left.\frac{\partial c_A}{\partial z}\right|_{z=0} = 0 \qquad t > 0 \quad (4\text{-}22)$$

$$D_{A(eff)}\left.\frac{\partial c_A}{\partial z}\right|_{z=L} + k_a\cdot c_A(z,t)\big|_{z=L} = 0 \qquad t > 0 \quad (4\text{-}23)$$

$$c_A(z,t)\big|_{t=0} = c_{A0} \qquad z\in[0,L] \quad (4\text{-}24)$$

This system is illustrated by the following diagram:

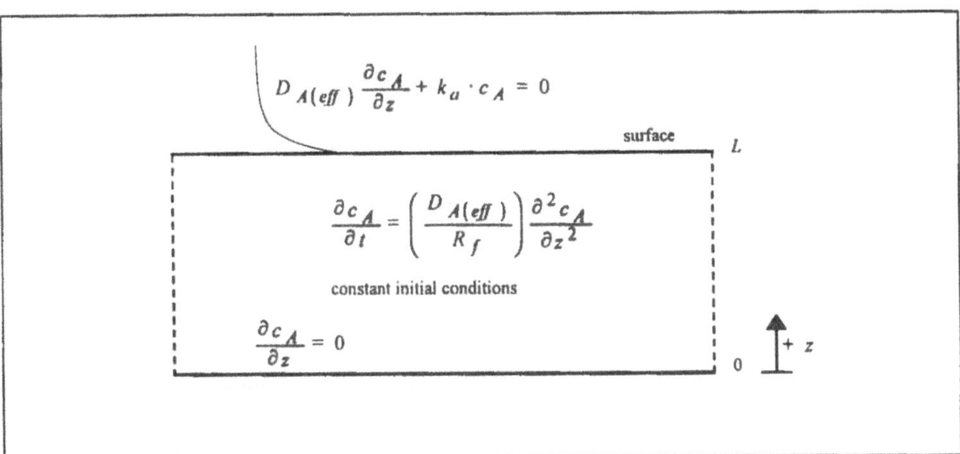

Figure 4-4 Diffusion in a finite layer with uniform initial concentration, mass transfer or reaction at the surface, and zero flux at the base.

The following expression is used for calculating the concentration

$$c_A(z,t) = \frac{2c_{A0}}{L}\sum_{n=1}^{\infty}\left\{\exp\left(-\frac{D_{A(eff)}}{R_f}\beta_n{}^2 t\right)\cdot\frac{1}{\beta_n}\cos(\beta_n z)\sin(\beta_n z)\right\} \tag{4-25}$$

and calculating the surface flux

$$j_A(t)\Big|_{z=L} = \frac{2D_{A(eff)}c_{A0}}{L}\sum_{n=1}^{\infty}\left\{\exp\left(-\frac{D_{A(eff)}}{R_f}\beta_n{}^2 t\right)\cdot\sin^2(\beta_n L)\right\} \tag{4-26}$$

where β_n is given by the positive roots of

$$\beta_n\tan(\beta_n L) = \left(\frac{k_a}{D_{A(eff)}}\right) \tag{4-27}$$

4.2.5 Case 5: Finite layer with arbitrary initial concentrations, zero concentration at the surface, zero flux at the base, and first-order decay

A system is defined by the dynamic equation (4-2) and with boundary and initial conditions the same as in case 1.

This system is illustrated by the following diagram:

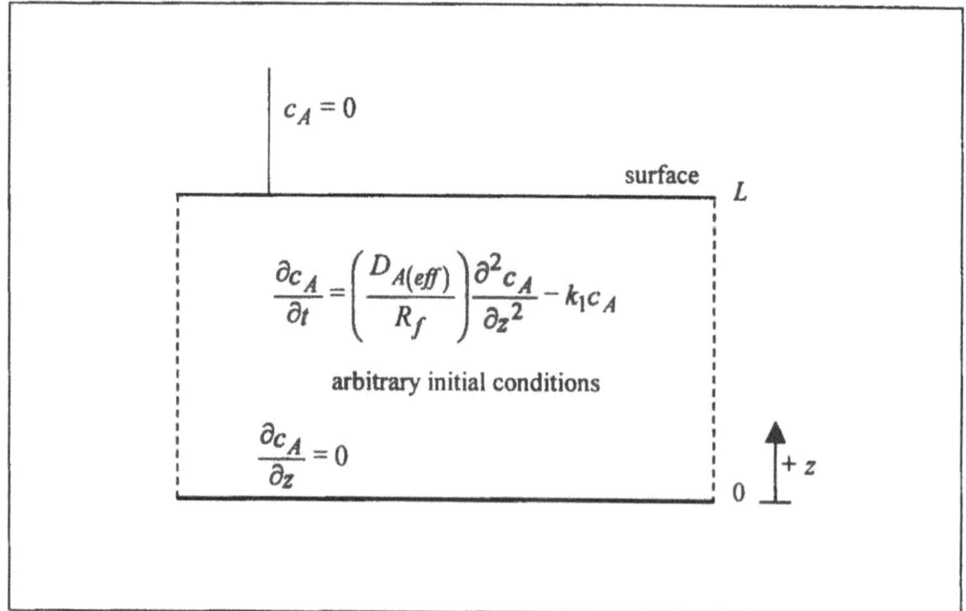

Figure 4-5 Diffusion in a finite layer with arbitrary initial concentrations, zero concentration at the surface, zero flux at the base, and first-order decay.

The following expression is used for calculating the concentration

$$c_A(z,t) = \frac{2}{L}\exp(-k_1 t)\sum_{n=1}^{\infty}\left\{\exp\left(-\frac{D_{A(eff)}}{R_f}\beta_n{}^2 t\right)\cdot\cos(\beta_n z)\cdot\int_0^L c_{A0}(z')\cdot\cos(\beta_n z')dz'\right\}$$

(4-28)

and calculating the surface flux

$$j_A(t)\big|_{z=L} = -\frac{2D_{A(eff)}}{L}\exp(-k_1 t)\sum_{n=1}^{\infty}\left\{\exp\left(-\frac{D_{A(eff)}}{R_f}\beta_n{}^2 t\right)\cdot\beta_n\cdot(-1)^n\cdot\int_0^L c_{A0}(z')\cdot\cos(\beta_n z')dz'\right\}$$

(4-29)

where β_n is given by

$$\beta_n = \frac{\pi}{L}\left(\frac{2n-1}{2}\right) \quad (4\text{-}30)$$

4.2.6 Case 6: Finite layer with uniform initial concentration, zero surface concentration, zero flux at the base, and first-order decay

A system is defined by the dynamic equation (4-2) and with boundary and initial conditions the same as in case 2.

This system is illustrated by the following diagram:

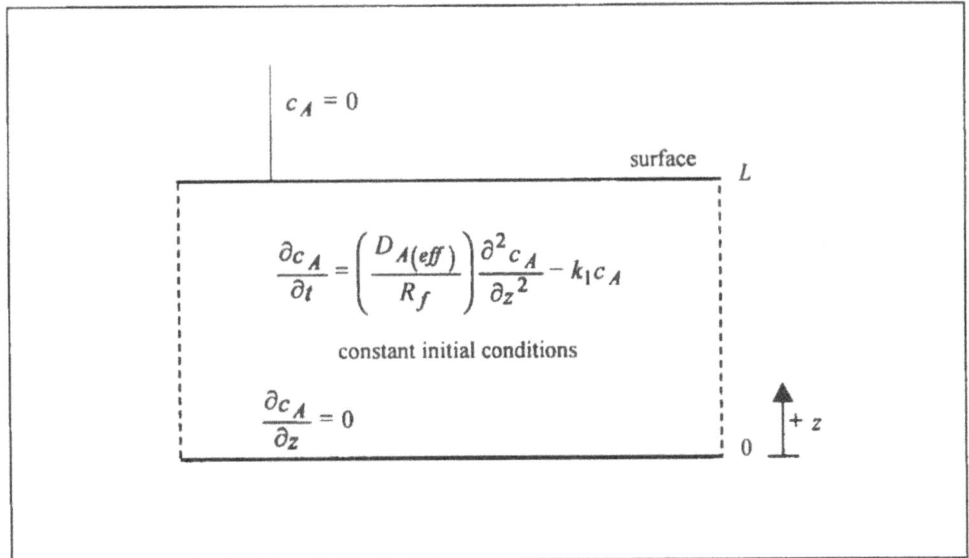

Figure 4-6 Diffusion in a finite layer with uniform initial concentrations, zero concentration at the surface, zero flux at the base, and first-order decay.

The following expression is used for calculating the concentration

$$c_A(z,t) = -\frac{2c_{A0}}{L}\exp(-k_1 t)\sum_{n=1}^{\infty}\left\{\exp\left(-\frac{D_{A(eff)}}{R_f}\beta_n^{\,2}t\right)\cdot\cos(\beta_n z)\cdot\frac{1}{\beta_n}\cdot(-1)^n\right\}$$

(4-31)

and calculating the surface flux

$$j_A(t)\big|_{z=L} = \frac{2D_{A(eff)}c_{A0}}{L}\exp(-k_1 t)\sum_{n=1}^{\infty}\exp\left(-\frac{D_{A(eff)}}{R_f}\beta_n^{\,2}t\right) \tag{4-32}$$

where β_n is given by

$$\beta_n = \frac{\pi}{L}\left(\frac{2n-1}{2}\right) \tag{4-33}$$

4.2.7 Case 7: Finite layer with arbitrary initial concentrations, mass transfer or reaction at the surface, zero flux at the base, and first-order decay

A system is defined by the dynamic equation (4-2) and with boundary and initial conditions the same as in case 3.

This system is illustrated by the following diagram:

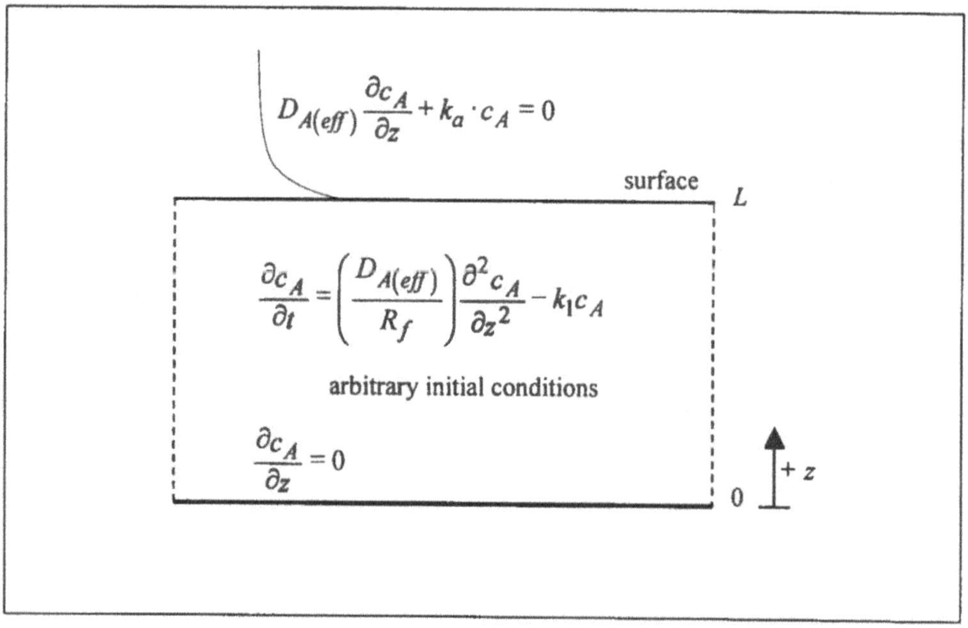

Figure 4-7 Diffusion in a finite layer with arbitrary initial concentrations, mass transfer or reaction at the surface, zero flux at the base, and first-order decay.

The following expression is used for calculating the concentration

$$c_A(z,t) = \frac{2}{L}\exp(-k_1 t)\sum_{n=1}^{\infty}\left\{\exp\left(-\frac{D_{A(eff)}}{R_f}\beta_n{}^2 t\right)\cdot\cos(\beta_n z)\cdot\int_0^L c_{A0}(z')\cdot\cos(\beta_n z')dz'\right\} \tag{4-34}$$

and calculating the surface flux

$$j_A(t)\big|_{z=L} = \frac{2D_{A(eff)}}{L}\exp(-k_1 t)\sum_{n=1}^{\infty}\left\{\exp\left(-\frac{D_{A(eff)}}{R_f}\beta_n{}^2 t\right)\cdot\beta_n\cdot\sin(\beta_n L)\cdot\int_0^L c_{A0}(z')\cdot\cos(\beta_n z')dz'\right\} \tag{4-35}$$

where β_n is given by the positive roots of

$$\beta_n \tan(\beta_n L) = \left(\frac{k_a}{D_{A(eff)}}\right) \tag{4-36}$$

4.2.8 Case 8: Finite layer with uniform initial concentration, mass transfer or reaction at the surface, zero flux at the base, and first-order decay

A system is defined by the dynamic equation (4-2) and with boundary and initial conditions the same as in case 2.

This system is illustrated by the following diagram:

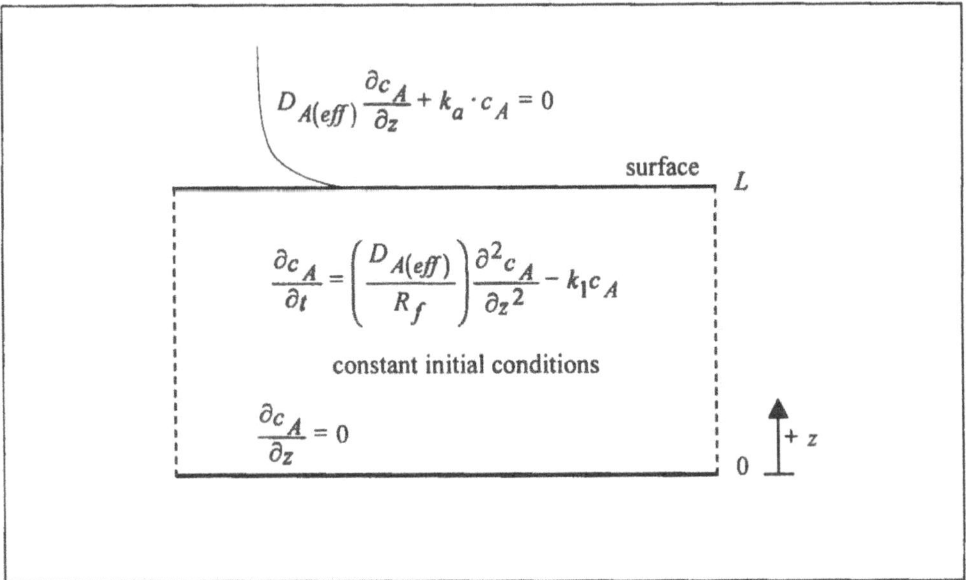

Figure 4-8 Diffusion in a finite layer with uniform initial concentration, mass transfer or reaction at the surface, zero flux at the base, and first-order decay.

The following expression is used for calculating the concentration

$$c_A(z,t)=\frac{2c_{A0}}{L}\exp(-k_1t)\sum_{n=1}^{\infty}\left\{\exp\left(-\frac{D_{A(eff)}}{R_f}\beta_n^{\,2}t\right)\cdot\frac{1}{\beta_n}\cos(\beta_n z)\sin(\beta_n z)\right\} \quad (4\text{-}37)$$

and calculating the surface flux

$$j\ (t)\big|_{z=L}=\frac{2D_{A(eff)}c_{A0}}{L}\exp(-k_1t)\sum_{n=1}^{\infty}\left\{\exp\left(-\frac{D_{A(eff)}}{R_f}\beta_n^{\,2}t\right)\cdot\sin^2(\beta_n L)\right\} \quad (4\text{-}38)$$

where β_n is given by the positive roots of

$$\beta_n\tan(\beta_n L)=\left(\frac{k_a}{D_{A(eff)}}\right) \quad (4\text{-}39)$$

4.3 Numerical evaluation

Both concentration and flux equations require the sum of an infinite series. Higher frequency components of the solution, due to the sine and cosine functions, are provided due to increasing values of n. However, the significance of these higher frequency terms diminishes with increasing n due to its presence in the exponential term.

For the numerical evaluation of the concentration and surface flux, the number of significant terms to be added, n_{max}, needs to be determined. Determining the number of significant terms will be discussed in Section 4.3.2 for the zero surface concentration cases, and 4.3.3 for the surface mass transfer boundary condition.

For the cases with arbitrary initial conditions an integral with the initial condition function, $c_{A0}(z)$, needs to be evaluated. A method of calculating this *initialization integral* is discussed in Section 4.3.1.

The convective boundary condition cases have the additional computational problem of finding roots to a transcendental function for its numerical evaluation. This is further discussed in Section 4.3.4.

4.3.1 Evaluation of the initial condition integral

For the cases with arbitrary initial conditions, cases 1, 3, 5, and 7, the initial concentration is given by

$$c_A(z,t)=c_{A0}(z) \qquad \text{at } t=0 \quad (4\text{-}40)$$

In these cases, an initial condition integral, or *initialization integral*, of the following form is required to be evaluated for a given value of β_n:

$$\int_0^L c_{A0}(z)\cdot\cos(\beta_n z)dz \tag{4-41}$$

The numerical procedure used to evaluate this integral depends on how the form in which the initial conditions, $c_{A0}(z)$, are given. For example, if the initial condition is of a simple functional form, the integral may be evaluated analytically.

More often, the initial conditions are specified as discrete values. The diagram below shows one method of specifying the initial conditions in such a manner, by a series of discrete constant blocks.

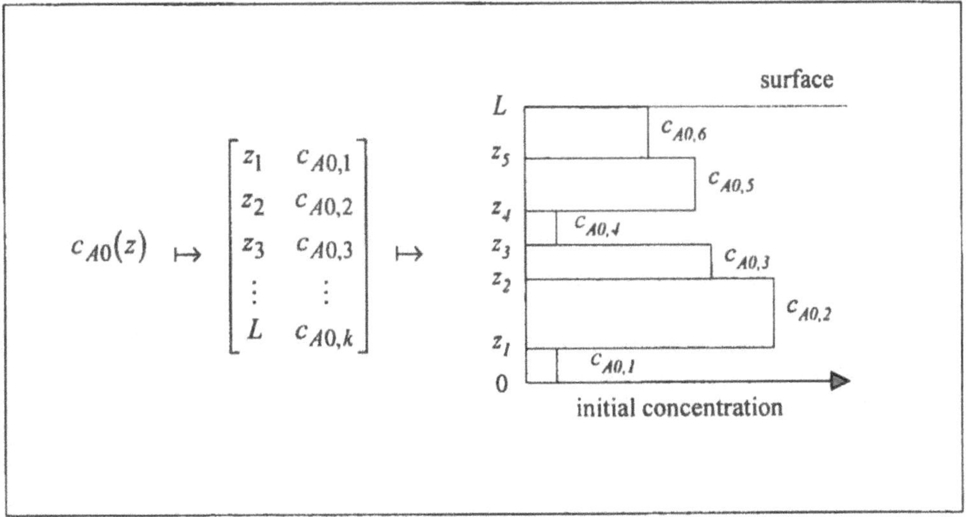

Figure 4-9 Method of specifying arbitrary initial conditions for a finite layer system.

The initialization integral is then evaluated by the following expression:

$$\int_0^L c_{A0}(z)\cdot\cos(\beta_n z)dz = c_{A0,1}\int_0^{z_1}\cos(\beta_n z)dz + c_{A0,2}\int_{z_1}^{z_2}\cos(\beta_n z)dz + \ldots$$

$$\ldots + c_{A0,k}\int_{z_{k-1}}^{L}\cos(\beta_n z)dz$$

$$= \frac{1}{\beta_n}\left\{\begin{matrix} c_{A0,1}\left[\sin(\beta_n z)\right]_0^{z_1} + c_{A0,2}\left[\sin(\beta_n z)\right]_{z_1}^{z_2} + \ldots \\ \ldots + c_{A0,k}\left[\sin(\beta_n z)\right]_{z_{k-1}}^{L} \end{matrix}\right\}$$

$$= \frac{1}{\beta_n}\left\{\begin{matrix} \left(c_{A0,1} - c_{A0,2}\right)\sin(\beta_n z_1) + \left(c_{A0,2} - c_{A0,3}\right)\sin(\beta_n z_2) + \ldots \\ \ldots + \left(c_{A0,k-1} - c_{A0,k}\right)\sin(\beta_n z_{k-1}) + c_{A0,k}\cdot\sin(\beta_n L) \end{matrix}\right\}$$

(4-42)

4.3.2 Cases 1 and 2: zero surface concentration

Both the concentration and flux equations require the calculation of an infinite series, i.e., $n = 1,2,3\ldots$. However, as the negative square of n is present in the exponential term, the significance of each term decreases with increasing n. A higher number of terms is required for predicting results at short time intervals. Truncation of the series without a sufficient number of terms often gives rise to oscillations in the predicted solution.

Due to the ease in which the terms are generated, a fixed arbitrary number may be set for the whole calculation range, $n_{max} = 20$ say. This value is increased if oscillations are still present in the resulting profile.

Hence numerically, the concentration for, say, case 2 is calculated with

$$c_A(z,t) = -\frac{2c_{A0}}{L}\sum_{n=1}^{n_{max}}\left\{\exp\left(-\frac{D_{A(eff)}}{R_f}\beta_n{}^2 t\right)\cdot\cos(\beta_n z)\cdot\frac{1}{\beta_n}\cdot(-1)^n\right\} \tag{4-43}$$

Similarly the surface flux is calculated summing contributions up to n_{max}.

4.3.3 Cases 3 and 4: surface mass transfer

Both the concentration and flux equations require the calculation of an infinite number of eigenvalues, $\beta_n : n = 1,2,3\ldots$ However as the negative square of the eigenvalue is present in the exponential term, the significance of each eigenvalue decreases with increasing n.

A range of significant eigenvalues, $\beta_n \in (0, \beta_{max}]$, needs to be determined in order to numerically evaluate the concentration and flux. Initially estimate β_{max} to give approximately 20 to 30 eigenvalues. That is

$$\beta_{max} = \frac{(2n_{max} - 1)\pi}{2L} \qquad n_{max} \sim 20 - 30 \qquad (4\text{-}44)$$

Hence numerically, the concentration for, say, case 3 is calculated with

$$c_A(z,t) = \frac{2}{L}\sum_{n=1}^{n_{max}}\left\{\exp\left(-\frac{D_{A(eff)}}{R_f}\beta_n{}^2 t\right)\cdot\cos(\beta_n z)\cdot\int_0^L c_{A0}(z')\cdot\cos(\beta_n z')dz'\right\} \quad (4\text{-}45)$$

where n_{max} is the number of roots in the range $\beta_n \in (0, \beta_{max}]$.

Similarly the surface flux is calculated summing contributions up to n_{max}.

4.3.4 Determining transcendental function roots

Rewriting the transcendental function as a tangent equated to a hyperbola,

$$\tan(\beta_n L) = \left(\frac{k_a}{D_{A(eff)}}\right)\frac{1}{\beta_n} \qquad (4\text{-}46)$$

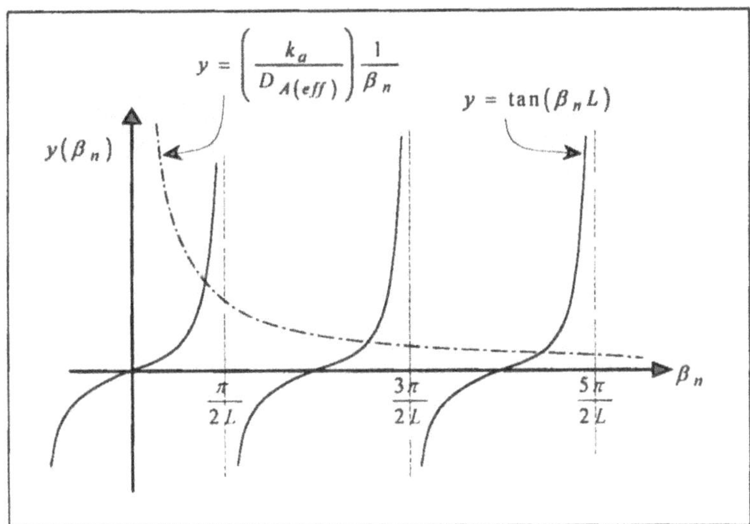

Figure 4-10 Roots to the transcendental function for a finite layer with a mass transfer boundary condition.

We can clearly see in the diagram above the intersection of the all the solutions to the equation are bracketed in the following manner:

$$\beta_1 \in \left(0, \frac{\pi}{2L}\right), \beta_2 \in \left(\frac{\pi}{2L}, \frac{3\pi}{2L}\right), \beta_3 \in \left(\frac{3\pi}{2L}, \frac{5\pi}{2L}\right), \beta_4 \in \left(\frac{5\pi}{2L}, \frac{7\pi}{2L}\right)\ldots$$

A simple bisection algorithm is used to find the roots to the transcendental function. Although a number of more efficient approaches can be used to find the roots, the bisection can be reliably applied to all of the root-finding needs in this volume without fears of missing roots. Asymptotes are present in both these given upper and lower bounds. Further details regarding the bisection algorithm is given in Appendix C.

4.4 Development

4.4.1 Separation of variables

Consider diffusion through a finite medium given by equation (4-1). Define the apparent diffusion (or retarded diffusion) as

$$\alpha = \left(\frac{D_{A(eff)}}{R_f} \right) \tag{4-47}$$

The partial differential equation is solved using a separation of variables technique. By the principle of the technique, the concentration is assumed to be separable into space- and time-dependant functions in the form

$$c_A(z,t) = \Psi(z) \cdot \Gamma(t) \tag{4-48}$$

For cases 1 to 4, substituting this into the dynamic equation (4-1) we obtain

$$\frac{1}{\Psi(z)} \frac{d^2\Psi}{dz^2} = \frac{1}{\alpha \Gamma(t)} \frac{d\Gamma}{dt} \tag{4-49}$$

In this equation, the left-hand side is a function of the space variable, z, alone and the right-hand side of the time variable, t, alone. The only way this equality holds is if both sides are equal to the same constant. Say the constant chosen is $-\beta^2$, thus we have

$$\frac{1}{\Psi(z)} \frac{d^2\Psi}{dz^2} = \frac{1}{\alpha \Gamma(t)} \frac{d\Gamma}{dt} = -\beta^2 \tag{4-50}$$

Discussion of cases 5 to 8 dynamic equation (4-2) is deferred to Section 4.4.4.

4.4.2 Solution to the temporal problem

The temporal (time-dependent) part of the system satisfies the differential equation

$$\frac{d\Gamma}{dt} + \alpha\beta^2\Gamma(t) = 0 \tag{4-51}$$

A possible solution of the above equation is

$$\Gamma(t) = \exp\left[-\alpha\beta^2 t\right] \tag{4-52}$$

4.4.3 Solution to the spatial problem

The spatial (space-dependent) part of the system satisfies the differential equation

$$\frac{d^2\Psi}{dz^2} + \beta^2\Psi(z) = 0 \qquad z \in [0, L] \tag{4-53}$$

The remaining portion of the spatial problem is given by the boundary conditions. For example, for case 1, the boundary conditions given by equations (4-4) and (4-5) are transformed to

$$\frac{d\Psi}{dz} = 0 \qquad \text{at } z = 0 \tag{4-54}$$

$$\Psi(z) = 0 \qquad \text{at } z = L \tag{4-55}$$

This auxiliary problem formed by the spatial part is an *eigenvalue problem*, as solutions only exist for certain values of the separation parameter, β. The values of $\beta = \beta_n, n = 1,2,3,...$ in which solutions can exist are called the ***eigenvalues*** of the system. The corresponding solutions, $\Psi(\beta_n, z)$, are called the *eigenfunctions*.

When β is not an eigenvalue the problem has only a trivial solution, that is, $\Psi(\beta, z) = 0$. The complete solution to the system, $c_A(z,t)$ is constructed by a linear superposition of the above-separated elementary solutions in the form

$$\begin{aligned} c_A(z,t) &= \sum_{n=1}^{\infty} \delta_n \cdot \Psi(\beta_n, z) \cdot \Gamma(\beta_n, t) \\ &= \sum_{n=1}^{\infty} \delta_n \cdot \Psi(\beta_n, z) \exp\left[-\alpha \beta_n^{\ 2} t\right] \end{aligned} \tag{4-56}$$

where δ_n are the constant coefficients used to satisfy the initial boundary conditions. The values of δ_n can be explicitly determined by solving equation (4-56) at $t = 0$ and using the property that the eigenfunctions are orthogonal.

$$c_A(z,t)\big|_{t=0} = \sum_{n=1}^{\infty} \delta_n \cdot \Psi(\beta_n, z) \qquad z \in [0, L] \tag{4-57}$$

$$\int_0^L \Psi(\beta_m, z) \cdot \Psi(\beta_n, z)\, dz = \begin{cases} 0 & \text{for } m \neq n \\ N(\beta_n) & \text{for } m = n \end{cases} \tag{4-58}$$

where the *normalization integral* (or the *norm*), $N(\beta_n)$, is defined as

$$N(\beta_n) = \int_0^L \left[\Psi(\beta_n, z)\right]^2 dz \tag{4-59}$$

Operating on both sides of equation (4-57) with the operator $\int_0^L \Psi(\beta_n, z)\, dz$ and utilizing the orthogonality property given in equation (4-58), we have

$$\delta_n = \frac{1}{N(\beta_n)} \int_0^L \Psi(\beta_n, z) \cdot c_A(z,0)\, dz \qquad (4\text{-}60)$$

Substituting this into equation (4-56), we obtain the general solution

$$c_A(z,t) = \sum_{n=1}^{\infty} \exp\left[-\alpha \beta_n^{\,2} t\right] \frac{1}{N(\beta_n)} \Psi(\beta_n, z) \int_0^L \Psi(\beta_n, z') \cdot c_A(z',0)\, dz' \qquad (4\text{-}61)$$

Differentiating equation (4-61) at $z = L$ gives a general solution for the surface flux

$$j_A(t)\Big|_{z=L} = -D_{A(eff)} \frac{\partial c_A}{\partial z}\bigg|_{z=L}$$

$$= -D_{A(eff)} \sum_{n=1}^{\infty} \left\{ \exp\left(-\frac{D_{A(eff)}}{R_f} \beta_n^{\,2} t \right) \cdot \frac{1}{N(\beta_n)} \cdot \frac{\partial \Psi}{\partial z}\bigg|_{z=L} \cdot \int_0^L c_{A0}(z') \cdot \Psi(\beta_n, z')\, dz' \right\}$$

(4-62)

4.4.3.1 *Case 1 and 2 eigenvalue problem*

The spatial boundary conditions for cases 1 and 2 are the same. These give rise to the following eigenvalue problem.

$$\frac{d^2\Psi}{dz^2} + \beta^2 \Psi(z) = 0 \qquad z \in [0, L] \qquad (4\text{-}63)$$

$$\frac{d\Psi}{dz} = 0 \qquad \text{at } z = 0 \qquad (4\text{-}64)$$

$$\Psi(z) = 0 \qquad \text{at } z = L \qquad (4\text{-}65)$$

The solution of this auxiliary problem is given by

$$\Psi(\beta_n, z) = \cos(\beta_n z) \qquad (4\text{-}66)$$

where β_n is the positive roots of

$$\cos(\beta_n L) = 0 \qquad (4\text{-}67)$$

Hence the following relations result:

$$\frac{1}{N(\beta_n)} = \frac{2}{L} \qquad (4\text{-}68)$$

$$\int_0^L \Psi(\beta_n, z)\, dz = \frac{1}{\beta_n} \sin(\beta_n L) \qquad (4\text{-}69)$$

$$\left.\frac{\partial \Psi}{\partial z}\right|_{z=L} = -\beta_n \sin(\beta_n L) \tag{4-70}$$

The eigenvalues, β_n, can be simplified to

$$\beta_n = \frac{\pi}{L}\left(\frac{2n-1}{2}\right) \tag{4-71}$$

thus,

$$\sin(\beta_n L) = -(-1)^n \tag{4-72}$$

Substituting the above solution of the eigenvalue problem into the generalized solution for arbitrary initial conditions (i.e., case 1), we obtain the following expression for the concentration:

$$c_A(z,t) = \frac{2}{L}\sum_{n=1}^{\infty}\left\{\exp\left(-\frac{D_{A(eff)}}{R_f}\beta_n{}^2 t\right)\cdot\cos(\beta_n z)\cdot\int_0^L c_{A0}(z')\cdot\cos(\beta_n, z')dz'\right\} \tag{4-73}$$

and, surface flux

$$\left.j_A(t)\right|_{z=L} = -\frac{2D_{A(eff)}}{L}\sum_{n=1}^{\infty}\left\{\exp\left(-\frac{D_{A(eff)}}{R_f}\beta_n{}^2 t\right)\cdot\beta_n\cdot(-1)^n\cdot\int_0^L c_{A0}(z')\cdot\cos(\beta_n, z')dz'\right\} \tag{4-74}$$

where β_n is given by equation (4-71).

For case 2, the initial concentration profile is constant, $c_{A0}(z') = c_{A0}$. Thus substituting this into equations (4-73) and (4-74) we obtain

$$c_A(z,t) = -\frac{2c_{A0}}{L}\sum_{n=1}^{\infty}\left\{\exp\left(-\frac{D_{A(eff)}}{R_f}\beta_n{}^2 t\right)\cdot\cos(\beta_n z)\cdot\frac{1}{\beta_n}\cdot(-1)^n\right\} \tag{4-75}$$

and

$$\left.\dot{n}(t)\right|_{z=L} = \frac{2D_{A(eff)}c_{A0}}{L}\sum_{n=1}^{\infty}\exp\left(-\frac{D_{A(eff)}}{R_f}\beta_n{}^2 t\right) \tag{4-76}$$

4.4.3.2 *Case 3 and 4 eigenvalue problem*

The spatial boundary conditions for cases 3 and 4 are the same. These give rise to the following eigenvalue problem.

$$\frac{d^2\Psi}{dz^2} + \beta^2\Psi(z) = 0 \qquad z \in [0, L] \tag{4-77}$$

$$\frac{d\Psi}{dz} = 0 \qquad \text{at } z = 0 \tag{4-78}$$

$$D_{A(eff)} \frac{\partial \Psi}{\partial z} + k_a \cdot \Psi(z) = 0 \qquad \text{at } z = L \tag{4-79}$$

The solution of this auxiliary problem is given by

$$\Psi(\beta_n, z) = \cos(\beta_n z) \tag{4-80}$$

where β_n is the positive roots of

$$\beta_n \tan(\beta_n L) = \left(\frac{k_a}{D_{A(eff)}} \right) \tag{4-81}$$

Hence the following relations result:

$$\frac{1}{N(\beta_n)} = \frac{2\beta_n}{\cos(\beta_n L)\sin(\beta_n L) + \beta_n L} \tag{4-82}$$

$$\int_0^L \Psi(\beta_n, z) dz = \frac{1}{\beta_n} \sin(\beta_n L) \tag{4-83}$$

$$\left. \frac{\partial \Psi}{\partial z} \right|_{z=L} = -\beta_n \sin(\beta_n L) \tag{4-84}$$

Substituting the above solution of the eigenvalue problem into the generalized solution, we obtain the following expression for the concentration

$$c_A(z,t) = \frac{2}{L} \sum_{n=1}^{\infty} \left\{ \exp\left(-\frac{D_{A(eff)}}{R_f} \beta_n^{\,2} t \right) \cdot \cos(\beta_n z) \cdot \int_0^L c_{A0}(z') \cdot \cos(\beta_n z') dz' \right\} \tag{4-85}$$

and surface flux

$$j_A(t)\big|_{z=L} = \frac{2D_{A(eff)}}{L} \sum_{n=1}^{\infty} \left\{ \exp\left(-\frac{D_{A(eff)}}{R_f} \beta_n^{\,2} t \right) \cdot \beta_n \cdot \sin(\beta_n L) \cdot \int_0^L c_{A0}(z') \cdot \cos(\beta_n z') dz' \right\} \tag{4-86}$$

where β_n is given by the positive roots of equation (4-81).

For case 4, the initial concentration profile is constant, $c_{A0}(z') = c_{A0}$. Thus substituting this into equations (4-85) and (4-86) we obtain

$$c_A(z,t) = \frac{2c_{A0}}{L} \sum_{n=1}^{\infty} \left\{ \exp\left(-\frac{D_{A(eff)}}{R_f} \beta_n^{\,2} t \right) \cdot \frac{1}{\beta_n} \cos(\beta_n z) \sin(\beta_n z) \right\} \tag{4-87}$$

and

$$j_A(t)\Big|_{z=L} = \frac{2D_{A(eff)}}{L}\sum_{n=1}^{\infty}\left\{\exp\left(-\frac{D_{A(eff)}}{R_f}\beta_n{}^2 t\right)\cdot\sin^2(\beta_n L)\right\} \tag{4-88}$$

4.4.4 Variable transformation for first-order decay

Solving the system with the inclusion of a first-order decay of the contaminant species, as described by the dynamic equation (4-2), we use the following variable substitution:

$$c_A(z,t) = \hat{c}(z,t)\exp(-k_1 t) \tag{4-89}$$

Effect of transformation on the dynamic equation

The transformed left-hand side of equation (4-2), the partial derivative with respect to time gives

$$\frac{\partial c_A}{\partial t} = -k_1\hat{c}(z,t)\exp(-k_1 t) + \frac{\partial \hat{c}}{\partial t}\exp(-k_1 t) \tag{4-90}$$

The transformed right-hand side, spatial terms give

$$\left(\frac{D_{A(eff)}}{R_f}\right)\frac{\partial^2 c_A}{\partial z^2} - k_1 c_A(z,t) = \left(\frac{D_{A(eff)}}{R_f}\right)\frac{\partial^2 \hat{c}}{\partial z^2}\exp(-k_1 t) - k_1\hat{c}(z,t)\exp(-k_1 t) \tag{4-91}$$

Hence, the transformed dynamic equations result in

$$\therefore \quad \frac{\partial \hat{c}}{\partial t} = \left(\frac{D_{A(eff)}}{R_f}\right)\frac{\partial^2 \hat{c}}{\partial z^2} \tag{4-92}$$

The net effect of the transformation is to remove the reaction term from the dynamic equation, resulting in the standard diffusion-adsorption formulation.

Effect of transformation on initial conditions

At time equal to zero the transformed initial concentration results in

$$c_A(z,0) = \hat{c}(z,0)\exp(0) \equiv \hat{c}(z,0) \tag{4-93}$$

Hence, all initial conditions remain unaltered after the transformation.

Effect of the transformation on boundary conditions

Transformation of the bottom boundary results in

$$\frac{\partial \hat{c}}{\partial z}\exp(-k_1 t) = 0 \qquad \text{at } z = 0,\ t > 0 \tag{4-94}$$

Transformation of the upper boundary condition for the case of zero surface concentration gives

$$\hat{c}(z,t)\exp(-k_1 t) = 0 \qquad \text{at } z = L,\ t > 0 \quad (4\text{-}95)$$

and for a mass transfer upper boundary gives

$$D_{A(eff)}\frac{\partial \hat{c}}{\partial z}\exp(-k_1 t) + k_a \cdot \hat{c}(z,t)\exp(-k_1 t) = 0 \qquad \text{at } z = L,\ t > 0 \quad (4\text{-}96)$$

Dividing all these equations by the common factor $\exp(-k_1 t)$ shows that the boundary conditions remain unaltered after the transformation.

Effect on the solution and surface flux

As the transformed system of equations results in the standard formulation with diffusion and adsorption mechanisms only, we solve for $\hat{c}(z,t)$ as described in the previous Section 4.4 for cases 1 to 4. The resulting solution then is inverted back into terms of $c_A(z,t)$ by equation (4-89).

The net effect of the decay term is to include an additional exponential time term to the solution. Since the flux is determined by a differential with respect to space, it too is only modified by this additional factor.

References

Özisk, M.N. (1993) *Heat Conduction*, 2nd ed., John Wiley & Sons, New York.

Carslaw, H.S., Jaeger, J.C. (1959) *Conduction of Heat in Solids*, Clarendon Press, Oxford.

5 Diffusion in a two-layer composite system

5.1 Introduction

Multilayer soil/sediment bed models are used in systems in which zones of distinctly different properties are present. Examples of such systems include soil columns with layering due to natural sedimentation, contaminated soil capped by clean earthen material, systems with a bioturbated top zone, and those with a combination of dry and damp zones.

The solutions giving the mobile phase concentration profile and surface flux rate for several two finite layer systems are presented. Cases where surface mass transfer and contaminant decay are significant and insignificant are both analyzed. These solutions are summarized in Section 5.2. A brief discussion on the numerical evaluation of these solutions is given in Section 5.3. The system equations were solved using a separation of variables technique. The development of the solutions is outlined in Section 5.4.

5.2 Analysis summary

5.2.1 System dynamics and general solution for a two-layer composite

The dynamics of a contaminant species A due to diffusion and adsorption processes, in a finite soil/sediment layer $z \in [0, b]$ which contains two distinct zones, is defined by the following equations:

$$\frac{\partial c_{A,1}}{\partial t} = \left(\frac{D_{A(eff),1}}{R_{f,1}}\right)\frac{\partial^2 c_{A,1}}{\partial z^2} \qquad z \in [0, a] \tag{5-1}$$

$$\frac{\partial c_{A,2}}{\partial t} = \left(\frac{D_{A(eff),2}}{R_{f,2}}\right)\frac{\partial^2 c_{A,2}}{\partial z^2} \qquad z \in [a, b] \tag{5-2}$$

where $c_A(z,t)$ is the concentration of species A in the mobile pore air or pore water phase and $z = a$ is the location of the interface between the two layers. Perfect contacting is assumed at the interface, hence the concentration at that position is the identical and no accumulation occurs. Mathematically this is expressed as

$$c_{A,1}(z,t) = c_{A,2}(z,t) \qquad \text{at } z = a,\ t > 0 \tag{5-3}$$

$$D_{A(eff),1}\frac{\partial c_{A,1}}{\partial z} = D_{A(eff),2}\frac{\partial c_{A,2}}{\partial z} \qquad \text{at } z = a,\ t > 0 \tag{5-4}$$

Surface flux from the top interface, $z = b$, is given by

$$j_A(t)\Big|_{z=b} = -D_{A(eff),2}\frac{\partial c_{A,2}}{\partial z}\bigg|_{z=b} \tag{5-5}$$

The general solution for the mobile phase concentration with the dynamics determined by equations (5-1) and (5-2) is given by

$$c_{A,i}(z,t)=\sum_{n=1}^{\infty}\left\{\exp\left(-\beta_n{}^2 t\right)\cdot\frac{1}{N(\beta_n)}\cdot\Psi_i(\beta_n,z)\cdot I_0(\beta_n)\right\} \tag{5-6}$$

and surface flux from the top interface, $z = b$, is given by

$$j_A(t)\Big|_{z=b}=-D_{A(eff),2}\sum_{n=1}^{\infty}\left\{\exp\left(-\beta_n{}^2 t\right)\cdot\frac{1}{N(\beta_n)}\cdot\left.\frac{\partial\Psi_2}{\partial z}\right|_{z=b}\cdot I_0(\beta_n)\right\} \tag{5-7}$$

where $\Psi_i(\beta_n,z)$ and β_n are the eigenfunctions and eigenvalues of the system. The parameters that define these eigenfunctions and eigenvalues are determined by the system boundary conditions. Further discussion and the general form of the eigenfunctions are given in Section 5.2.2. Specific solutions for several different boundary conditions are given in Sections 5.2.3 and 5.2.4.

For the above general solution, $N(\beta_n)$ is the *normalization integral* given by

$$N(\beta_n)=R_{f,1}\int_0^a\left[\Psi_1(\beta_n,z')\right]^2 dz'+R_{f,2}\int_a^b\left[\Psi_2(\beta_n,z')\right]^2 dz' \tag{5-8}$$

and $I_0(\beta_n)$ is the *initialization integral* given by

$$I_0(\beta_n)=R_{f,1}\int_0^a\Psi_1(\beta_n,z')\cdot c_{A0}(z')dz'+R_{f,2}\int_a^b\Psi_2(\beta_n,z')\cdot c_{A0}(z')dz' \tag{5-9}$$

where $c_{A0}(z)$ is the initial condition.

If a first-order decay mechanism of the contaminant is also of significance, the governing dynamic equations (5-1) and (5-2) are replaced by the following:

$$\frac{\partial c_{A,1}}{\partial t}=\left(\frac{D_{A(eff),1}}{R_{f,1}}\right)\frac{\partial^2 c_{A,1}}{\partial z^2}-k_1 c_{A,1}(z,t) \qquad z\in[0,a] \tag{5-10}$$

$$\frac{\partial c_{A,2}}{\partial t}=\left(\frac{D_{A(eff),2}}{R_{f,2}}\right)\frac{\partial^2 c_{A,2}}{\partial z^2}-k_1 c_{A,2}(z,t) \qquad z\in[a,b] \tag{5-11}$$

The general solution for the mobile phase concentration with the dynamics determined by equations (5-10) and (5-11) is given by

$$c_{A,i}(z,t)=\exp(-k_1 t)\sum_{n=1}^{\infty}\left\{\exp\left(-\beta_n{}^2 t\right)\cdot\frac{1}{N(\beta_n)}\cdot\Psi_i(\beta_n,z)\cdot I_0(\beta_n)\right\} \tag{5-12}$$

and surface flux from the top interface, $z = b$, is given by

$$j_A(t)\Big|_{z=b} = -D_{A(eff),2}\exp(-k_1 t)\sum_{n=1}^{\infty}\left\{\exp\left(-\beta_n^{\,2}t\right)\cdot\frac{1}{N(\beta_n)}\cdot\frac{\partial\Psi_2}{\partial z}\bigg|_{z=b}\cdot I_0(\beta_n)\right\} \quad (5\text{-}13)$$

where $\Psi_i(\beta_n,z)$ and β_n are the eigenfunctions and eigenvalues of the system and the normalization and initialization integral is defined in equations (5-8) and (5-9) above. The eigenfunctions and eigenvalues are defined in the same manner as that for the cases without the decay process dynamics.

5.2.2 System eigenfunctions and eigenvalues

The eigenfunctions are an orthogonal set of functions used to generate the overall concentration profile. The one-dimensional composite diffusion eigenfunctions have the general form of

$$\Psi_i(\beta_n,z) = A_{i,n}\sin\left(\sqrt{\frac{R_{f,i}}{D_{A(eff),i}}}\cdot\beta_n\cdot z\right) + B_{i,n}\cos\left(\sqrt{\frac{R_{f,i}}{D_{A(eff),i}}}\cdot\beta_n\cdot z\right) \quad (5\text{-}14)$$

where the eigenvalues, $\beta_n: n = 1,2,3\ldots$, and eigenfunction coefficients, $A_{i,n}$ and $B_{i,n}$, are determined by the system boundary conditions.

Following are forms of the eigenfunction that are required for the calculation for the normalization and initialization integrals and hence the concentration and surface flux. Define

$$\lambda_{i,n} = \beta_n\sqrt{\frac{R_{f,i}}{D_{A(eff),i}}} \quad (5\text{-}15)$$

The partial differential of the generalized eigenfunction with respect to z is

$$\frac{\partial\Psi_i}{\partial z} = \left\{\lambda_{i,n}A_{i,n}\right\}\cos\left(\lambda_{i,n}z\right) + \left\{-\lambda_{i,n}B_{i,n}\right\}\sin\left(\lambda_{i,n}z\right) \quad (5\text{-}16)$$

The indefinite integral of the generalized eigenfunction with respect to z is

$$\int\Psi(\beta_n,z)dz = \left\{-\frac{A_{i,n}}{\lambda_{i,n}}\right\}\cos\left(\lambda_{i,n}z\right) + \left\{\frac{B_{i,n}}{\lambda_{i,n}}\right\}\sin\left(\lambda_{i,n}z\right) \quad (5\text{-}17)$$

The square of the generalized eigenfunction is

$$\left[\Psi_i(\beta_n,z)\right]^2 = A_{i,n}^{\;2}\sin^2\left(\lambda_{i,n}z\right) + 2A_{i,n}B_{i,n}\sin\left(\lambda_{i,n}z\right)\cos\left(\lambda_{i,n}z\right) + B_{i,n}\cos^2\left(\lambda_{i,n}z\right) \quad (5\text{-}18)$$

The integral of the square of the generalized eigenfunction is

$$\int\left[\Psi_i(\beta_n,z)\right]^2 dz =$$

$$\frac{\left[A_{i,n}^{\;2}+B_{i,n}^{\;2}\right]\lambda_{i,n}z+\left[B_{i,n}^{\;2}-A_{i,n}^{\;2}\right]\cos\left(\lambda_{i,n}z\right)\sin\left(\lambda_{i,n}z\right)+2A_{i,n}B_{i,n}\sin^2\left(\lambda_{i,n}z\right)}{2\lambda_{i,n}} \tag{5-19}$$

5.2.3 Case 1: Two-layer finite system with arbitrary initial concentrations, zero concentration at the surface, and zero flux at the base

A system is defined with the following boundary conditions

$$\left.\frac{\partial c_{A,1}}{\partial z}\right|_{z=0}=0 \qquad t>0 \tag{5-20}$$

$$\left.c_{A,2}(z,t)\right|_{z=b}=0 \qquad t>0 \tag{5-21}$$

and initial conditions

$$\left.c_{A,1}(z,t)\right|_{t=0}=c_{A0}(z) \qquad z\in[0,a] \tag{5-22}$$

$$\left.c_{A,2}(z,t)\right|_{t=0}=c_{A0}(z) \qquad z\in[a,b] \tag{5-23}$$

The interface boundary conditions are given by equations (5-3) and (5-4).

With the dynamics given by equations (5-1) and (5-2), the system is illustrated by the following diagram:

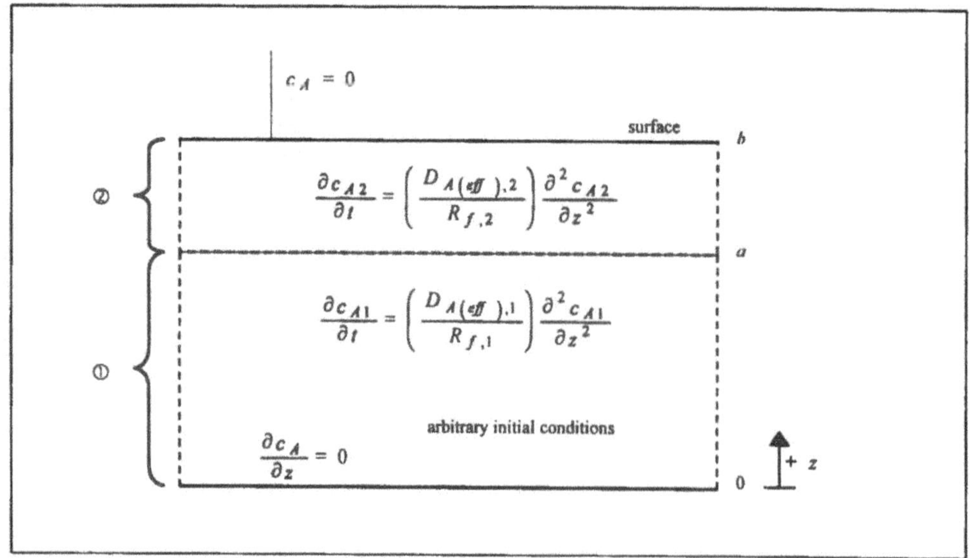

Figure 5-1 Diffusion in a two-layer finite system with arbitrary initial concentrations, zero concentration at the surface, and zero flux at the base.

With the dynamics given by equations (5-10) and (5-11), the system is illustrated by the following diagram:

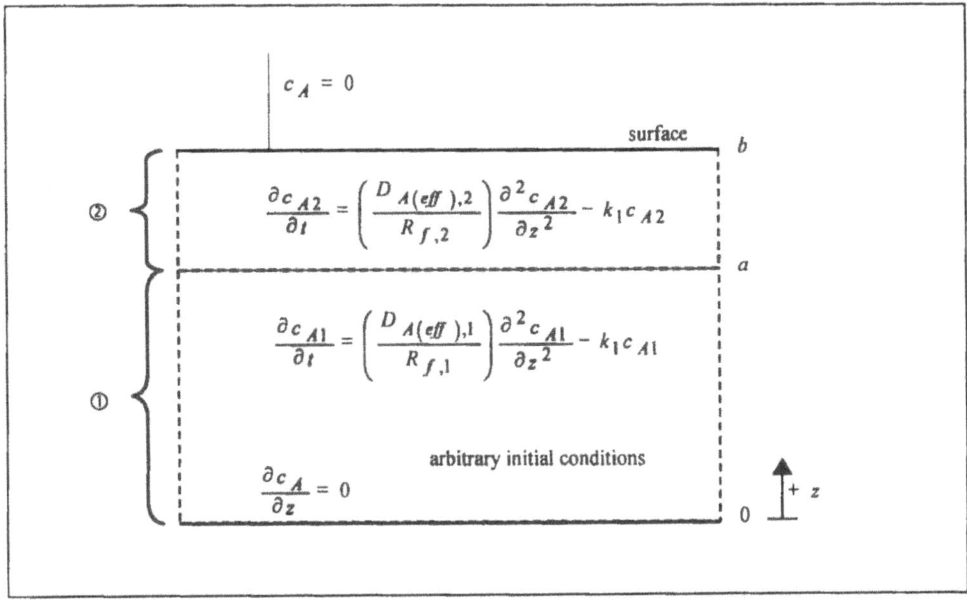

Figure 5-2 Diffusion in a two-layer finite system with arbitrary initial concentrations, zero concentration at the surface, zero flux at the base, and first-order decay.

In both dynamic situations the coefficients for the eigenfunctions are given by

$$A_{1,n} = 0 \tag{5-24}$$

$$B_{1,n} = 1 \tag{5-25}$$

$$A_{2,n} = \frac{\cos\left(\frac{\beta_n}{\sqrt{\alpha_1}}a\right)\cos\left(\frac{\beta_n}{\sqrt{\alpha_2}}b\right)}{\sin\left[\frac{\beta_n}{\sqrt{\alpha_2}}(a-b)\right]} \tag{5-26}$$

$$B_{2,n} = \frac{-\cos\left(\frac{\beta_n}{\sqrt{\alpha_1}}a\right)\sin\left(\frac{\beta_n}{\sqrt{\alpha_2}}b\right)}{\sin\left[\frac{\beta_n}{\sqrt{\alpha_2}}(a-b)\right]} \tag{5-27}$$

where $\alpha_i = \left(\frac{D_{A(eff),i}}{R_{f,i}}\right)$ and β_n is the positive roots of

$$\left(\frac{1-\sqrt{\frac{D_1 \cdot R_{f,1}}{D_2 \cdot R_{f,2}}}}{1+\sqrt{\frac{D_1 \cdot R_{f,1}}{D_2 \cdot R_{f,2}}}}\right)\cos\left[\left(\frac{a}{\sqrt{\alpha_1}}+\frac{a-b}{\sqrt{\alpha_2}}\right)\beta_n\right]+\cos\left[\left(\frac{a}{\sqrt{\alpha_1}}-\frac{a-b}{\sqrt{\alpha_2}}\right)\beta_n\right]=0 \quad (5\text{-}28)$$

These eigenvalues and eigenfunction coefficients are substituted into the system eigenfunction as given in Section 5.2.2 and are then used to evaluate the concentration or surface flux, as described in Section 5.2.1.

5.2.4 Case 2: Two-layer finite system with arbitrary initial concentrations, mass transfer or reaction at the surface, and zero flux at the base

A system is defined with the following boundary conditions

$$\left.\frac{\partial c_{A,1}}{\partial z}\right|_{z=0}=0 \qquad t>0 \qquad (5\text{-}29)$$

$$D_{A(eff),2}\left.\frac{\partial c_{A,2}}{\partial z}\right|_{z=b}+k_a \cdot c_{A,2}(z,t)\big|_{z=b}=0 \qquad t>0 \qquad (5\text{-}30)$$

and initial conditions

$$c_{A,1}(z,t)\big|_{t=0}=c_{A0}(z) \qquad z\in[0,a] \qquad (5\text{-}31)$$

$$c_{A,2}(z,t)\big|_{t=0}=c_{A0}(z) \qquad z\in[a,b] \qquad (5\text{-}32)$$

The interface boundary conditions are given by equations (5-3) and (5-4).

With the dynamics given by equations (5-1) and (5-2), the system is illustrated by the following diagram:

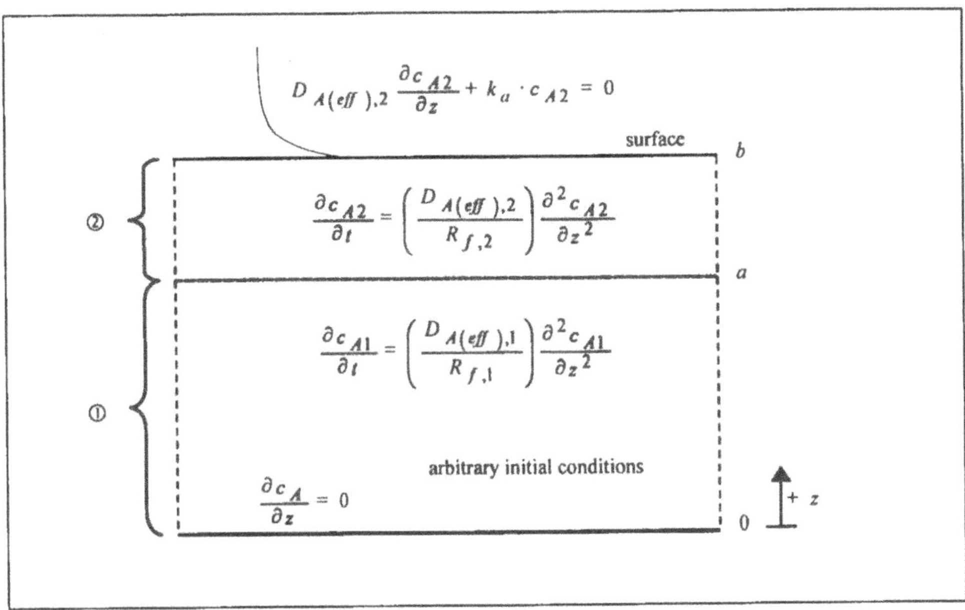

Figure 5-3 Diffusion in a two-layer finite system with arbitrary initial concentrations, mass transfer or reaction at the surface, and zero flux at the base.

With the dynamics given by equations (5-10) and (5-11), the system is illustrated by the following diagram:

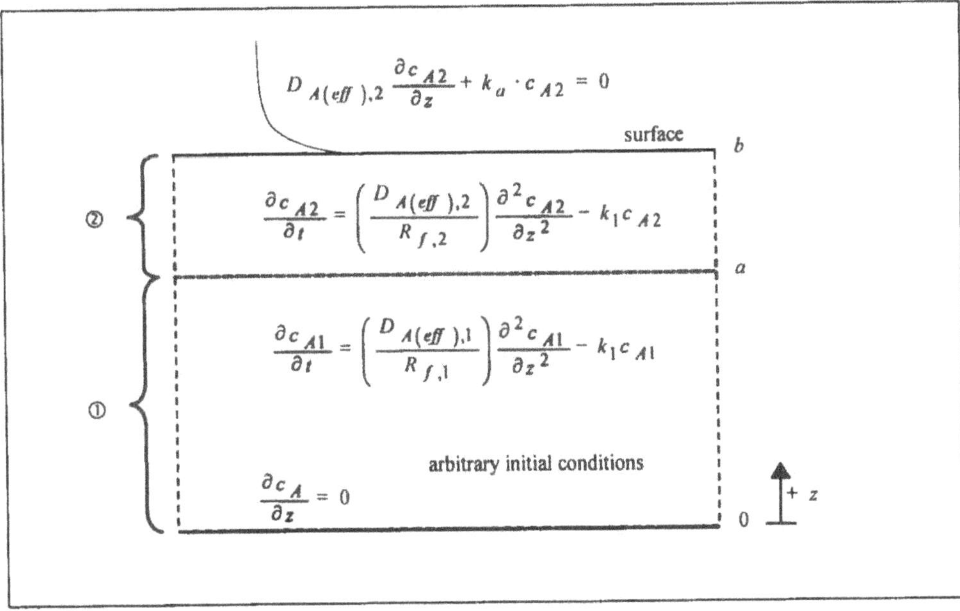

Figure 5-4 Diffusion in a two-layer finite system with arbitrary initial concentrations, mass transfer or reaction at the surface, zero flux at the base, and first-order decay.

In both dynamic situations the coefficients for the eigenfunctions are given by

$$A_{1,n} = 0 \tag{5-33}$$

$$B_{1,n} = 1 \tag{5-34}$$

$$A_{2,n} = -\frac{D_1}{D_2}\sqrt{\frac{\alpha_2}{\alpha_1}}\cos\left(\frac{\beta_n}{\sqrt{\alpha_2}}a\right)\sin\left(\frac{\beta_n}{\sqrt{\alpha_1}}a\right)+\cos\left(\frac{\beta_n}{\sqrt{\alpha_1}}a\right)\sin\left(\frac{\beta_n}{\sqrt{\alpha_2}}a\right) \tag{5-35}$$

$$B_{2,n} = \frac{D_1}{D_2}\sqrt{\frac{\alpha_2}{\alpha_1}}\sin\left(\frac{\beta_n}{\sqrt{\alpha_2}}a\right)\sin\left(\frac{\beta_n}{\sqrt{\alpha_1}}a\right)+\cos\left(\frac{\beta_n}{\sqrt{\alpha_2}}a\right)\cos\left(\frac{\beta_n}{\sqrt{\alpha_1}}a\right) \tag{5-36}$$

where $\alpha_i = \left(\frac{D_{A(eff),i}}{R_{f,i}}\right)$, and β_n is the positive roots of

$$\left[\left(\frac{D_1}{\sqrt{\alpha_1}}+\frac{D_2}{\sqrt{\alpha_2}}\right)\beta_n\right]\sin\left(\left[\frac{a}{\sqrt{\alpha_1}}+\frac{b-a}{\sqrt{\alpha_2}}\right]\beta_n\right)+\left[\left(\frac{D_1}{\sqrt{\alpha_1}}-\frac{D_2}{\sqrt{\alpha_2}}\right)\beta_n\right]\sin\left(\left[\frac{a}{\sqrt{\alpha_1}}-\frac{b-a}{\sqrt{\alpha_2}}\right]\beta_n\right)+\ldots$$

$$\ldots\left[-k_a\left(\frac{D_1}{D_2}\sqrt{\frac{\alpha_2}{\alpha_1}}+1\right)\right]\cos\left(\left[\frac{a}{\sqrt{\alpha_1}}+\frac{b-a}{\sqrt{\alpha_2}}\right]\beta_n\right)+\left[k_a\left(\frac{D_1}{D_2}\sqrt{\frac{\alpha_2}{\alpha_1}}-1\right)\right]\cos\left(\left[\frac{a}{\sqrt{\alpha_1}}-\frac{b-a}{\sqrt{\alpha_2}}\right]\beta_n\right)=0$$

(5-37)

These eigenvalues and eigenfunction coefficients are substituted into the system eigenfunction as given in Section 5.2.2 and are then used to evaluate the concentration or surface flux, as described in Section 5.2.1.

5.3 Numerical Evaluation

Both calculations for the concentration and surface flux require the sum of an infinite series of terms. As the terms tend to infinity, their effect on the solution decreases in significance. Numerically, the range in which the terms are still of significance needs to be determined. Following this, the eigenvalues in this range are evaluated. The eigenvalues are determined by finding roots to the appropriate transcendental functions determined by the system boundary conditions.

Once the eigenvalues have been determined, the eigenfunctions and hence the normalization integral, initialization integral, differential of the eigenfunction may be calculated. These are combined according to the general solution given in equations (5-6) and (5-7), or if the contaminant decay dynamics is also included, equations (5-12) and (5-13).

5.3.1 Concentration calculation

The following diagram shows the stages needed for the calculation of a mobile phase concentration at position $z \in [0,b]$ and time $t > 0$. Each box in the diagram below represents a stage in the calculation process. It has references to the appropriate equations or section in this manuscript.

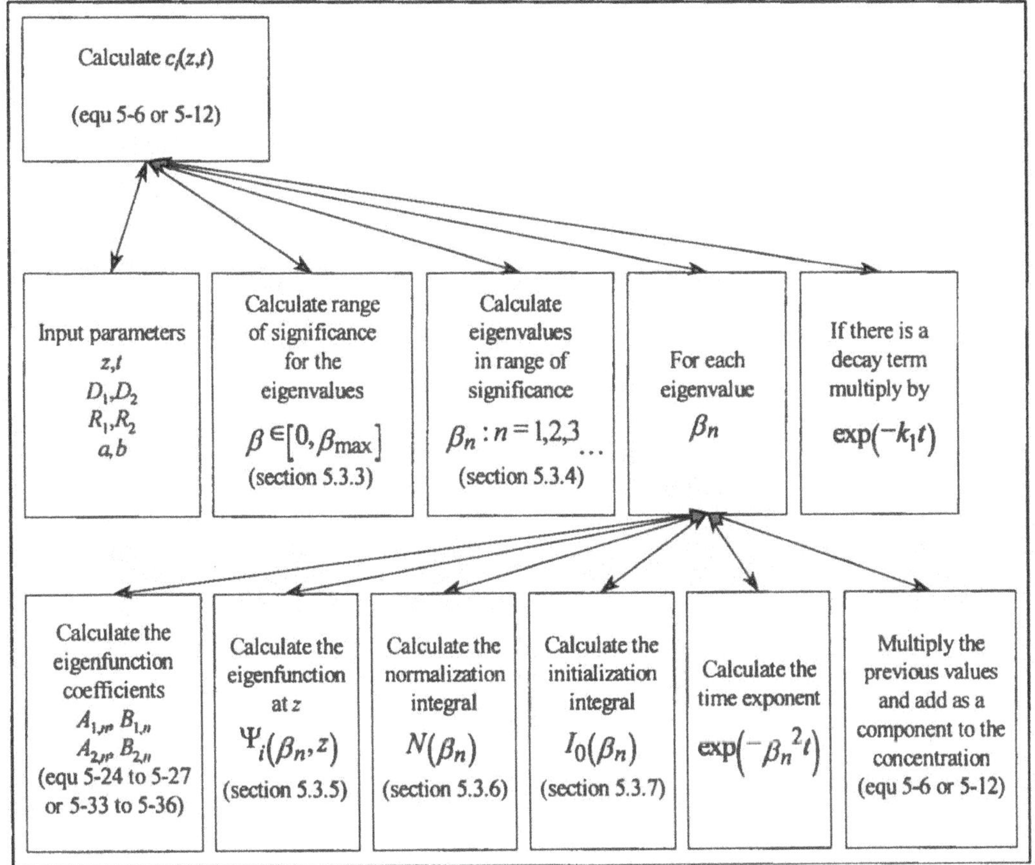

Figure 5-5 Information flow diagram for concentration calculation.

5.3.2 Surface flux calculation

The following diagram shows the stages needed for the calculation of the flux at the sediment surface at time $t > 0$. Each box in the diagram below represents a stage in the calculation process. It has references to the appropriate equations or section in this manuscript.

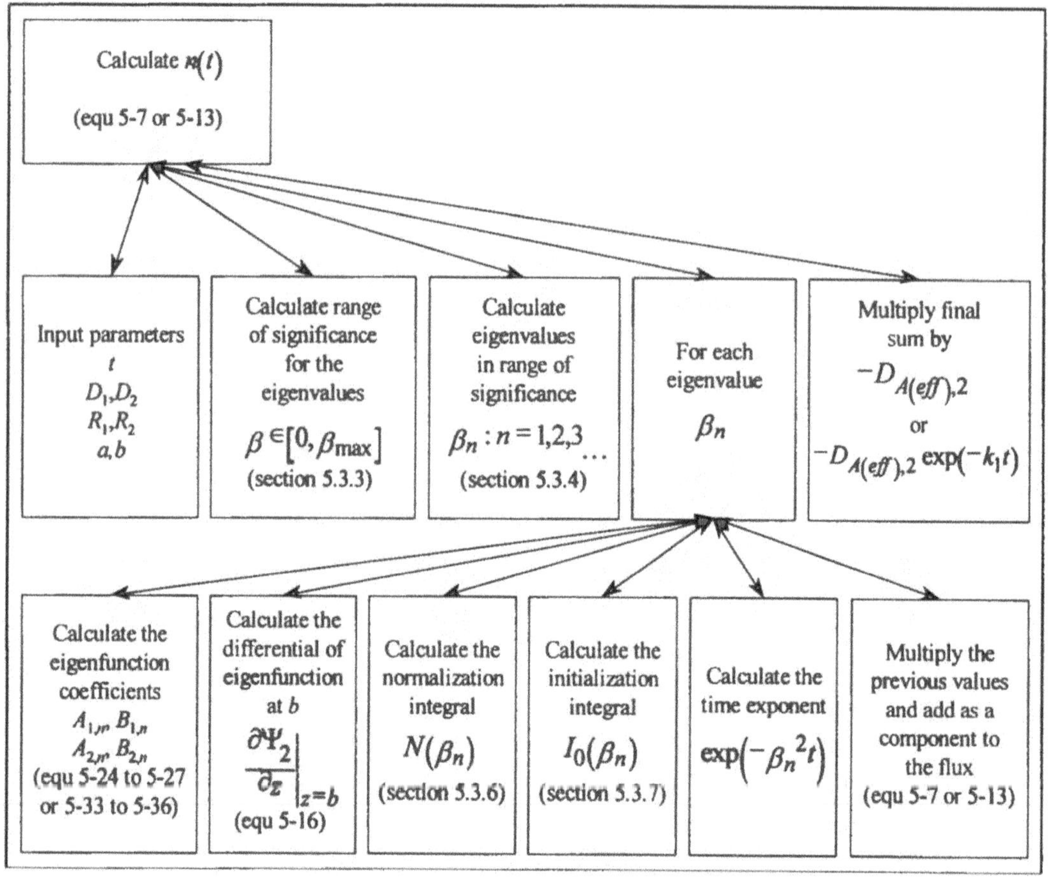

Figure 5-6 Information flow diagram for surface flux calculation.

5.3.3 Range of significance for eigenvalues

Both the concentration and flux equations require the calculation of an infinite number of eigenvalues, $\beta_n : n = 1,2,3\ldots$ However as the negative square of the eigenvalue is present in the exponential term, the significance of each eigenvalue decreases with increasing n.

A range of significant eigenvalues, $\beta_n \in (0, \beta_{max}]$, needs to be determined in order to numerically evaluate the concentration and flux. Generally, setting β_{max} equal to 20 to 30 times the period of the smallest coefficient in the transcendental function sine and cosine terms is sufficient. That is

$$\beta_{max} = \omega \frac{\pi}{\min\{\text{'sin' or 'cos' coefficients}\}} \qquad \omega \sim 20-30 \tag{5-38}$$

If more eigenvalues are needed to generate an acceptable solution increase the value of ω as appropriate.

5.3.4 Determination of eigenvalues in range

The eigenvalues are given for case 1 as the roots of the equation

$$C_1 \cos[C_2 \cdot \beta_n] + C_3 \cos[C_4 \cdot \beta_n] = 0 \tag{5-39}$$

or for case 2 as the roots of the equation

$$K_1 \cdot \beta_n \cdot \sin[K_2 \cdot \beta_n] + K_3 \cdot \beta_n \cdot \sin[K_4 \cdot \beta_n] + K_5 \cos[K_6 \cdot \beta_n] + K_7 \cos[K_8 \cdot \beta_n] = 0 \tag{5-40}$$

where the coefficients for case 1 are given by

$$C_1 = \left[\frac{1 - \sqrt{\dfrac{D_1 \cdot R_{f,1}}{D_2 \cdot R_{f,2}}}}{1 + \sqrt{\dfrac{D_1 \cdot R_{f,1}}{D_2 \cdot R_{f,2}}}} \right] \tag{5-41}$$

$$C_2 = \left[\frac{a}{\sqrt{\alpha_1}} + \frac{a-b}{\sqrt{\alpha_2}} \right] \tag{5-42}$$

$$C_3 = 1 \tag{5-43}$$

$$C_4 = \left[\frac{a}{\sqrt{\alpha_1}} - \frac{a-b}{\sqrt{\alpha_2}} \right] \tag{5-44}$$

or for case 2 are given by

$$K_1 = \left[\frac{D_1}{\sqrt{\alpha_1}} + \frac{D_2}{\sqrt{\alpha_2}} \right] \tag{5-45}$$

$$K_2 = \left[\frac{a}{\sqrt{\alpha_1}} + \frac{b-a}{\sqrt{\alpha_2}} \right] \tag{5-46}$$

$$K_3 = \left[\frac{D_1}{\sqrt{\alpha_1}} - \frac{D_2}{\sqrt{\alpha_2}} \right] \tag{5-47}$$

$$K_4 = \left[\frac{a}{\sqrt{\alpha_1}} - \frac{b-a}{\sqrt{\alpha_2}} \right] \tag{5-48}$$

$$K_5 = \left[k_a \left(\frac{D_1}{D_2} \sqrt{\frac{\alpha_2}{\alpha_1}} - 1 \right) \right] \tag{5-49}$$

$$K_6 = \left[\frac{a}{\sqrt{\alpha_1}} - \frac{b-a}{\sqrt{\alpha_2}}\right] \tag{5-50}$$

$$K_7 = \left[-k_a\left(\frac{D_1}{D_2}\sqrt{\frac{\alpha_2}{\alpha_1}}+1\right)\right] \tag{5-51}$$

$$K_8 = \left[\frac{a}{\sqrt{\alpha_1}} + \frac{b-a}{\sqrt{\alpha_2}}\right] \tag{5-52}$$

Roots of these equations are calculated within the range $\beta_n \in (0, \beta_{max}]$.

A rigorous bracketing procedure is required to capture all the roots within this range. Bracket boundaries are determined from the maxima and minima for each of the oscillatory functions. Each of these bracketed regions is checked to determine if a root is present. If a root does exist, a numerical search routine is used to determine its value.

Hence for case 1, the following two series of cosine maxima and minima are calculated:

$$\left\{0, \frac{\pi}{C_2}, \frac{2\pi}{C_2}, \frac{3\pi}{C_2}, \ldots\right\} \le \beta_{max} \quad, \left\{0, \frac{\pi}{C_4}, \frac{2\pi}{C_4}, \frac{3\pi}{C_4}, \ldots\right\} \le \beta_{max} \tag{5-53}$$

The above two series are then combined and sorted into a single series given by

$$\{b_1, b_2, b_3, b_4 \ldots\} \le \beta_{max} \qquad \text{where } b_i < b_{i+1} \text{ and } b_1 = 0 \tag{5-54}$$

Any redundant points are eliminated from the combined series. This series gives the search brackets to be tested.

Alternatively for case 2, the following four series of sine and cosine maxima and minima are calculated:

$$\left\{\frac{\pi}{2K_2}, \frac{3\pi}{2K_2}, \frac{5\pi}{2K_2}, \ldots\right\} \le \beta_{max} \quad, \left\{\frac{\pi}{2K_4}, \frac{3\pi}{2K_4}, \frac{5\pi}{2K_4}, \ldots\right\} \le \beta_{max} \quad,$$

$$\left\{0, \frac{\pi}{K_6}, \frac{2\pi}{K_6}, \frac{3\pi}{K_6}, \ldots\right\} \le \beta_{max} \quad, \left\{0, \frac{\pi}{K_8}, \frac{2\pi}{K_8}, \frac{3\pi}{K_8}, \ldots\right\} \le \beta_{max} \tag{5-55}$$

The above four series are then combined and sorted into a single series given by

$$\{b_1, b_2, b_3, b_4 \ldots\} \le \beta_{max} \qquad \text{where } b_i < b_{i+1} \text{ and } b_1 = 0 \tag{5-56}$$

Any redundant points are eliminated from the combined series. This series gives the search brackets to be tested.

Search brackets for the roots are given by the values of the sorted series. Defining the function

$$f(b) = \begin{cases} C_1 \cos[C_2 \cdot b] + C_3 \cdot \cos[C_4 \cdot b] & \text{case 1} \\ K_1 b \sin[C_2 \cdot b] + K_3 b \sin[K_4 \cdot b] + K_5 \cos[K_6 \cdot b] + K_7 \cos[K_8 \cdot b] & \text{case 2} \end{cases} \tag{5-57}$$

For each bracket the following condition is tested:

$$\text{if } f(b_i) \cdot f(b_{i+1}) \le 0 \text{ then root exists between } [b_i, b_{i+1}] \tag{5-58}$$

If the root exists on the lower bound, i.e., $f(b_i)=0$, the eigenvalue is recorded. We then proceed to the next bracket. If the root exists on the upper bound, i.e., $f(b_{i+1})=0$, the eigenvalue is not recorded. We then proceed to the next bracket where it will be picked up by the algorithm. If the root lies between the bracket bounds, a bisection algorithm is used to find the eigenvalue. This value is recorded.

This process is continued until all the bracketed regions in the range $(0,\beta_{max}]$ have been tested. The resulting series of recorded roots are the eigenvalues, β_n, in this range of significance.

Further details about the algorithms for bisection searches and for finding the roots of the stated transcendental problem can be found in Appendix C.

5.3.5 Eigenfunction evaluation

Given an eigenvalue, β_n, and system parameters, the coefficients for all the eigenfunctions can be evaluated from equations (5-24) to (5-27), for case 1, and from equations (5-22) to (5-36), for case 2.

Given a position, *z*, the layer in which it is located may be determined. The appropriate layer eigenfunction can then be evaluated from equation (5-14).

5.3.6 Normalization integral evaluation

Given an eigenvalue, β_n, and system parameters, the coefficients for the all the eigenfunctions can be evaluated from equations (5-24) to (5-27), for case 1, and from equations (5-33) to (5-36), for case 2.

The normalization integral can then be calculated from equations (5-8) and (5-19).

5.3.7 Initialization integral evaluation

The method of evaluating the initialization integral, given by equation (5-9), depends on how the initial concentration profile is given. If a functional form is available, $c_{A0}=f(z)$, the integral may be evaluated analytically. However, it is more common that the initial concentration is specified by a tabulation of discrete values.

If constant, discrete values are used, the initialization integral can be broken up into the sum of a series of definite integrals. For example, given the initial concentration distribution in the following diagram,

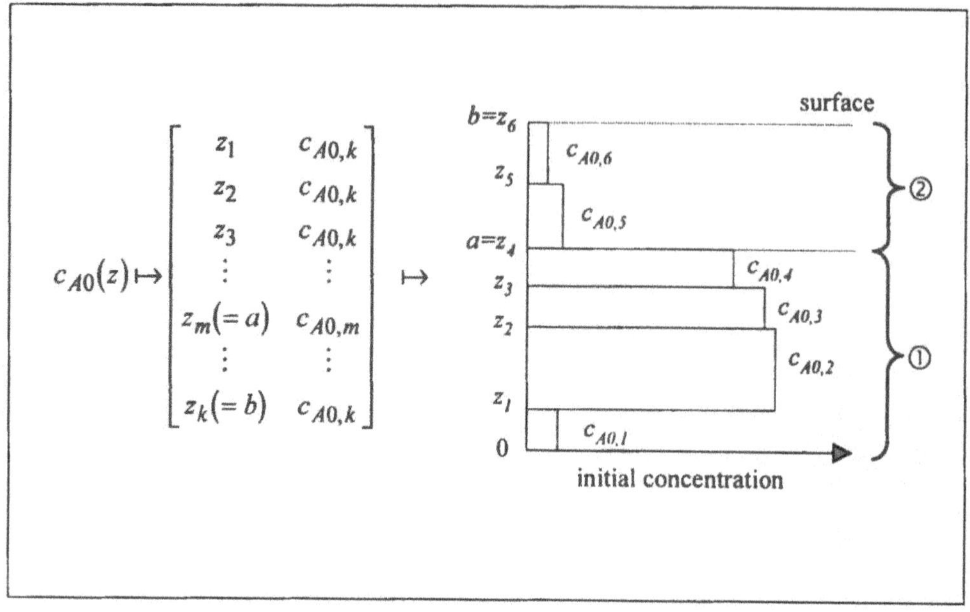

Figure 5-7 Method of specifying arbitrary initial conditions for a two-layer composite system.

the initialization integral can be calculated as

$$I_0(\beta_n) = R_{f,1}\left[\begin{array}{l} c_{A0,1}\int_0^{z_1}\Psi_1(\beta_n,z)dz + c_{A0,2}\int_{z_1}^{z_2}\Psi_1(\beta_n,z)dz \quad \ldots \\ \ldots + c_{A0,3}\int_{z_2}^{z_3}\Psi_1(\beta_n,z)dz + c_{A0,4}\int_{z_3}^{z_4}\Psi_1(\beta_n,z)dz \end{array}\right]\ldots$$
$$\ldots + R_{f,2}\left[c_{A0,5}\int_{z_4}^{z_5}\Psi_2(\beta_n,z)dz + c_{A0,6}\int_{z_5}^{z_6}\Psi_2(\beta_n,z)dz\right] \tag{5-59}$$

where the definite integral of the eigenfunction is given by equation (5-17).

5.4 Development

5.4.1 Separation of variables

The system of partial differential equations is solved using a separation of variables technique. By the principle of the technique, the concentration is assumed to be separable into independent functions of position and time, of the form

$$c_{A,i}(z,t) = \Psi_i(z) \cdot \Gamma(t) \tag{5-60}$$

Substituting this into the dynamic equations (5-1) and (5-2) we obtain

$$\alpha_i \frac{1}{\Psi_{i,n}(z)} \frac{d^2\Psi_i}{dz^2} = \frac{1}{\Gamma(t)} \frac{d\Gamma}{dt} \tag{5-61}$$

where $\alpha_i = \left(\frac{D_{A(eff),i}}{R_{f,i}} \right)$. Note discussion on the effect of the decay term and the dynamic equations (5-10) and (5-11) are deferred to Section 5.4.5.

In equation (5-61) the left-hand side is a function of the space variable, z, alone and the right-hand side of the time variable, t, alone. The only way this equality holds is if both sides are equal to the same constant. Say the constant chosen is $-\beta^2$, thus we have

$$\alpha_i \frac{1}{\Psi_i(z)} \frac{d^2\Psi_i}{dz^2} = \frac{1}{\Gamma(t)} \frac{d\Gamma}{dt} = -\beta^2 \tag{5-62}$$

β is termed the separation factor.

5.4.2 Solution to the temporal problem

The temporal part of equation (5-61) is given by

$$\frac{d\Gamma}{dt} = -\beta^2 \cdot \Gamma(t) \tag{5-63}$$

A possible solution for this equation is

$$\Gamma(t) = \exp\left(-\beta^2 t\right) \tag{5-64}$$

5.4.3 Solution to the spatial problem

The spatial equations generated from equation (5-61), and the specific boundary conditions gives rise to eigenvalue problems. The solution to each case is outlined below.

5.4.3.1 Case 1 eigenvalue problem

The eigenvalue problem can be stated as

$$\frac{d^2\Psi_{1,n}}{dz^2}+\frac{\beta_n{}^2}{\alpha_1}\Psi_{1,n}(z)=0 \qquad \text{for } z\in[0,a] \tag{5-65}$$

$$\frac{d^2\Psi_{2,n}}{dz^2}+\frac{\beta_n{}^2}{\alpha_2}\Psi_{2,n}(z)=0 \qquad \text{for } z\in[a,b] \tag{5-66}$$

with transformed boundary conditions of

$$\frac{d\Psi_1}{dz}=0 \qquad \text{at } z=0 \tag{5-67}$$

$$D_1\frac{d\Psi_1}{dz}=D_2\frac{d\Psi_2}{dz} \qquad \text{at } z=a \tag{5-68}$$

$$\Psi_1(z)=\Psi_2(z) \qquad \text{at } z=a \tag{5-69}$$

$$\Psi_2(z)=0 \qquad \text{at } z=b \tag{5-70}$$

Assuming the general form of

$$\Psi_i(\beta_n,z)=A_{i,n}\sin\left(\frac{\beta_n}{\sqrt{\alpha_i}}\cdot z\right)+B_{i,n}\cos\left(\frac{\beta_n}{\sqrt{\alpha_i}}\cdot z\right) \tag{5-71}$$

hence

$$\frac{d\Psi_i}{dz}=\left\{\frac{\beta_n}{\sqrt{\alpha_i}}\cdot A_{i,n}\right\}\cos\left(\frac{\beta_n}{\sqrt{\alpha_i}}\cdot z\right)+\left\{-\frac{\beta_n}{\sqrt{\alpha_i}}\cdot B_{i,n}\right\}\sin\left(\frac{\beta_n}{\sqrt{\alpha_i}}\cdot z\right) \tag{5-72}$$

Solving for the coefficients using the transformed boundary conditions

$$\begin{bmatrix} \frac{\beta_n}{\sqrt{\alpha_1}} & 0 & 0 & 0 \\ D_1\frac{\beta_n}{\sqrt{\alpha_1}}\cos\left(\frac{\beta_n}{\sqrt{\alpha_1}}a\right) & -D_1\frac{\beta_n}{\sqrt{\alpha_1}}\sin\left(\frac{\beta_n}{\sqrt{\alpha_1}}a\right) & -D_2\frac{\beta_n}{\sqrt{\alpha_2}}\cos\left(\frac{\beta_n}{\sqrt{\alpha_2}}a\right) & D_2\frac{\beta_n}{\sqrt{\alpha_2}}\sin\left(\frac{\beta_n}{\sqrt{\alpha_2}}a\right) \\ \sin\left(\frac{\beta_n}{\sqrt{\alpha_1}}a\right) & \cos\left(\frac{\beta_n}{\sqrt{\alpha_1}}a\right) & -\sin\left(\frac{\beta_n}{\sqrt{\alpha_2}}a\right) & -\cos\left(\frac{\beta_n}{\sqrt{\alpha_2}}a\right) \\ 0 & 0 & \sin\left(\frac{\beta_n}{\sqrt{\alpha_2}}b\right) & \cos\left(\frac{\beta_n}{\sqrt{\alpha_2}}b\right) \end{bmatrix}\begin{bmatrix} A_{1,n} \\ B_{1,n} \\ A_{2,n} \\ B_{2,n} \end{bmatrix}=\begin{bmatrix} 0 \\ 0 \\ 0 \\ 0 \end{bmatrix} \tag{5-73}$$

From the above matrix equation it can be seen that $A_{1,n}=0$. Thus $B_{1,n}$ may be set to any arbitrary constant. Let $B_{1,n}=1$, equation (5-74) becomes

$$\begin{bmatrix} -\sin\left(\frac{\beta_n}{\sqrt{\alpha_2}}a\right) & -\cos\left(\frac{\beta_n}{\sqrt{\alpha_2}}a\right) \\ \sin\left(\frac{\beta_n}{\sqrt{\alpha_2}}b\right) & \cos\left(\frac{\beta_n}{\sqrt{\alpha_2}}b\right) \end{bmatrix}\begin{bmatrix} A_{2,n} \\ B_{2,n} \end{bmatrix}=\begin{bmatrix} -\cos\left(\frac{\beta_n}{\sqrt{\alpha_1}}a\right) \\ 0 \end{bmatrix} \tag{5-74}$$

Hence

$$A_{2,n}=\frac{\cos\left(\frac{\beta_n}{\sqrt{\alpha_1}}a\right)\cos\left(\frac{\beta_n}{\sqrt{\alpha_2}}b\right)}{\sin\left(\frac{\beta_n}{\sqrt{\alpha_2}}a\right)\cos\left(\frac{\beta_n}{\sqrt{\alpha_2}}b\right)-\sin\left(\frac{\beta_n}{\sqrt{\alpha_2}}b\right)\cos\left(\frac{\beta_n}{\sqrt{\alpha_2}}a\right)} \tag{5-75}$$

$$B_{2,n}=\frac{-\cos\left(\frac{\beta_n}{\sqrt{\alpha_1}}a\right)\sin\left(\frac{\beta_n}{\sqrt{\alpha_2}}b\right)}{\sin\left(\frac{\beta_n}{\sqrt{\alpha_2}}a\right)\cos\left(\frac{\beta_n}{\sqrt{\alpha_2}}b\right)-\sin\left(\frac{\beta_n}{\sqrt{\alpha_2}}b\right)\cos\left(\frac{\beta_n}{\sqrt{\alpha_2}}a\right)} \tag{5-76}$$

These simply give coefficients of

$$A_{1,n}=0 \tag{5-77}$$

$$B_{1,n}=1 \tag{5-78}$$

$$A_{2,n}=\frac{\cos\left(\frac{\beta_n}{\sqrt{\alpha_1}}a\right)\cos\left(\frac{\beta_n}{\sqrt{\alpha_2}}b\right)}{\sin\left[\frac{\beta_n}{\sqrt{\alpha_2}}(a-b)\right]} \tag{5-79}$$

$$B_{2,n}=\frac{-\cos\left(\frac{\beta_n}{\sqrt{\alpha_1}}a\right)\sin\left(\frac{\beta_n}{\sqrt{\alpha_2}}b\right)}{\sin\left[\frac{\beta_n}{\sqrt{\alpha_2}}(a-b)\right]} \tag{5-80}$$

Eigenvalues, β_n, are calculated from the characteristic equation of the matrix in equation (5-73):

$$\det\begin{vmatrix}\frac{\beta_n}{\sqrt{\alpha_1}} & 0 & 0 & 0\\ D_1\frac{\beta_n}{\sqrt{\alpha_1}}\cos\left(\frac{\beta_n}{\sqrt{\alpha_1}}a\right) & -D_1\frac{\beta_n}{\sqrt{\alpha_1}}\sin\left(\frac{\beta_n}{\sqrt{\alpha_1}}a\right) & -D_1\frac{\beta_n}{\sqrt{\alpha_2}}\cos\left(\frac{\beta_n}{\sqrt{\alpha_2}}a\right) & D_1\frac{\beta_n}{\sqrt{\alpha_2}}\sin\left(\frac{\beta_n}{\sqrt{\alpha_2}}a\right)\\ \sin\left(\frac{\beta_n}{\sqrt{\alpha_1}}a\right) & \cos\left(\frac{\beta_n}{\sqrt{\alpha_1}}a\right) & -\sin\left(\frac{\beta_n}{\sqrt{\alpha_2}}a\right) & -\cos\left(\frac{\beta_n}{\sqrt{\alpha_2}}a\right)\\ 0 & 0 & \sin\left(\frac{\beta_n}{\sqrt{\alpha_2}}b\right) & \cos\left(\frac{\beta_n}{\sqrt{\alpha_2}}b\right)\end{vmatrix}=0 \tag{5-81}$$

This simplifies to

$$\left(\frac{1-\frac{D_1}{D_2}\sqrt{\frac{\alpha_2}{\alpha_1}}}{1+\frac{D_1}{D_2}\sqrt{\frac{\alpha_2}{\alpha_1}}}\right)\cos\left[\left(\frac{a}{\sqrt{\alpha_1}}+\frac{a-b}{\sqrt{\alpha_2}}\right)\beta_n\right]+\cos\left[\left(\frac{a}{\sqrt{\alpha_1}}-\frac{a-b}{\sqrt{\alpha_2}}\right)\beta_n\right]=0 \qquad (5\text{-}82)$$

Thus, the eigenvalues for the boundary conditions of case 1 are given by the positive roots of the above equation.

5.4.3.2 Case 2 eigenvalue problem

The eigenvalue problem can be stated as

$$\frac{d^2\Psi_{1,n}}{dz^2}+\frac{\beta_n^{\ 2}}{\alpha_1}\Psi_{1,n}(z)=0 \qquad \text{for } z\in[0,a] \qquad (5\text{-}83)$$

$$\frac{d^2\Psi_{2,n}}{dz^2}+\frac{\beta_n^{\ 2}}{\alpha_2}\Psi_{2,n}(z)=0 \qquad \text{for } z\in[a,b] \qquad (5\text{-}84)$$

with transformed boundary conditions of

$$\frac{d\Psi_1}{dz}=0 \qquad \text{at } z=0 \qquad (5\text{-}85)$$

$$D_1\frac{d\Psi_1}{dz}=D_2\frac{d\Psi_2}{dz} \qquad \text{at } z=a \qquad (5\text{-}86)$$

$$\Psi_1(z)=\Psi_2(z) \qquad \text{at } z=a \qquad (5\text{-}87)$$

$$\frac{d\Psi_2}{dz}+\left(\frac{k_a}{D_{A(eff),2}}\right)\Psi_2(z)=0 \qquad \text{at } z=b \qquad (5\text{-}88)$$

Assuming the general form of

$$\Psi_i(\beta_n,z)=A_{i,n}\sin\left(\frac{\beta_n}{\sqrt{\alpha_i}}\cdot z\right)+B_{i,n}\cos\left(\frac{\beta_n}{\sqrt{\alpha_i}}\cdot z\right) \qquad (5\text{-}89)$$

hence

$$\frac{d\Psi_i}{dz}=\left\{\frac{\beta_n}{\sqrt{\alpha_i}}\cdot A_{i,n}\right\}\cos\left(\frac{\beta_n}{\sqrt{\alpha_i}}\cdot z\right)+\left\{-\frac{\beta_n}{\sqrt{\alpha_i}}\cdot B_{i,n}\right\}\sin\left(\frac{\beta_n}{\sqrt{\alpha_i}}\cdot z\right) \qquad (5\text{-}90)$$

Solving for the coefficients using the transformed boundary conditions

$$\begin{bmatrix} \frac{\beta_n}{\sqrt{\alpha_1}} & 0 & 0 & 0 \\ D_1\frac{\beta_n}{\sqrt{\alpha_1}}\cos\left(\frac{\beta_n}{\sqrt{\alpha_1}}a\right) & -D_1\frac{\beta_n}{\sqrt{\alpha_1}}\sin\left(\frac{\beta_n}{\sqrt{\alpha_1}}a\right) & -D_2\frac{\beta_n}{\sqrt{\alpha_2}}\cos\left(\frac{\beta_n}{\sqrt{\alpha_2}}a\right) & D_2\frac{\beta_n}{\sqrt{\alpha_2}}\sin\left(\frac{\beta_n}{\sqrt{\alpha_2}}a\right) \\ \sin\left(\frac{\beta_n}{\sqrt{\alpha_1}}a\right) & \cos\left(\frac{\beta_n}{\sqrt{\alpha_1}}a\right) & -\sin\left(\frac{\beta_n}{\sqrt{\alpha_2}}a\right) & -\cos\left(\frac{\beta_n}{\sqrt{\alpha_2}}a\right) \\ 0 & 0 & \chi & \xi \end{bmatrix} \begin{bmatrix} A_{1,n} \\ B_{1,n} \\ A_{2,n} \\ B_{2,n} \end{bmatrix} = \begin{bmatrix} 0 \\ 0 \\ 0 \\ 0 \end{bmatrix} \tag{5-91}$$

where

$$\chi = \frac{\beta_n}{\sqrt{\alpha_2}}\cos\left(\frac{\beta_n}{\sqrt{\alpha_2}}b\right) + \left(\frac{k_a}{D_{A(eff),2}}\right)\sin\left(\frac{\beta_n}{\sqrt{\alpha_2}}b\right) \tag{5-92}$$

and

$$\xi = -\frac{\beta_n}{\sqrt{\alpha_2}}\sin\left(\frac{\beta_n}{\sqrt{\alpha_2}}b\right) + \left(\frac{k_a}{D_{A(eff),2}}\right)\cos\left(\frac{\beta_n}{\sqrt{\alpha_2}}b\right) \tag{5-93}$$

From the above matrix equation it can be seen that $A_{1,n} = 0$. Thus $B_{1,n}$ may be set to any arbitrary constant. Let $B_{1,n} = 1$, thus equation (5-91) becomes

$$\begin{bmatrix} -\sin\left(\frac{\beta_n}{\sqrt{\alpha_2}}a\right) & -\cos\left(\frac{\beta_n}{\sqrt{\alpha_2}}a\right) \\ -D_2\frac{\beta_n}{\sqrt{\alpha_2}}\cos\left(\frac{\beta_n}{\sqrt{\alpha_2}}a\right) & D_2\frac{\beta_n}{\sqrt{\alpha_2}}\sin\left(\frac{\beta_n}{\sqrt{\alpha_2}}a\right) \end{bmatrix} \begin{bmatrix} A_{2,n} \\ B_{2,n} \end{bmatrix} = \begin{bmatrix} -\cos\left(\frac{\beta_n}{\sqrt{\alpha_1}}a\right) \\ D_1\frac{\beta_n}{\sqrt{\alpha_1}}\sin\left(\frac{\beta_n}{\sqrt{\alpha_1}}a\right) \end{bmatrix} \tag{5-94}$$

Hence

$$A_{2,n} = -\frac{D_1}{D_2}\sqrt{\frac{\alpha_2}{\alpha_1}}\cos\left(\frac{\beta_n}{\sqrt{\alpha_2}}a\right)\sin\left(\frac{\beta_n}{\sqrt{\alpha_1}}a\right) + \cos\left(\frac{\beta_n}{\sqrt{\alpha_1}}a\right)\sin\left(\frac{\beta_n}{\sqrt{\alpha_2}}a\right) \tag{5-95}$$

$$B_{2,n} = \frac{D_1}{D_2}\sqrt{\frac{\alpha_2}{\alpha_1}}\sin\left(\frac{\beta_n}{\sqrt{\alpha_2}}a\right)\sin\left(\frac{\beta_n}{\sqrt{\alpha_1}}a\right) + \cos\left(\frac{\beta_n}{\sqrt{\alpha_2}}a\right)\cos\left(\frac{\beta_n}{\sqrt{\alpha_1}}a\right) \tag{5-96}$$

Eigenvalues, β_n, are calculated from the characteristic equation of the matrix in equation (5-91):

$$\det\begin{vmatrix} \frac{\beta_n}{\sqrt{\alpha_1}} & 0 & 0 & 0 \\ D_1\frac{\beta_n}{\sqrt{\alpha_1}}\cos\left(\frac{\beta_n}{\sqrt{\alpha_1}}a\right) & -D_1\frac{\beta_n}{\sqrt{\alpha_1}}\sin\left(\frac{\beta_n}{\sqrt{\alpha_1}}a\right) & -D_1\frac{\beta_n}{\sqrt{\alpha_2}}\cos\left(\frac{\beta_n}{\sqrt{\alpha_2}}a\right) & D_1\frac{\beta_n}{\sqrt{\alpha_2}}\sin\left(\frac{\beta_n}{\sqrt{\alpha_2}}a\right) \\ \sin\left(\frac{\beta_n}{\sqrt{\alpha_1}}a\right) & \cos\left(\frac{\beta_n}{\sqrt{\alpha_1}}a\right) & -\sin\left(\frac{\beta_n}{\sqrt{\alpha_2}}a\right) & -\cos\left(\frac{\beta_n}{\sqrt{\alpha_2}}a\right) \\ 0 & 0 & \chi & \xi \end{vmatrix} = 0 \tag{5-97}$$

This simplifies to

$$\left[\left(\frac{D_1}{\sqrt{\alpha_1}}+\frac{D_2}{\sqrt{\alpha_2}}\right)\beta_n\right]\sin\left(\left[\frac{a}{\sqrt{\alpha_1}}+\frac{b-a}{\sqrt{\alpha_2}}\right]\beta_n\right)+\left[\left(\frac{D_1}{\sqrt{\alpha_1}}-\frac{D_2}{\sqrt{\alpha_2}}\right)\beta_n\right]\sin\left(\left[\frac{a}{\sqrt{\alpha_1}}-\frac{b-a}{\sqrt{\alpha_2}}\right]\beta_n\right)+\ldots$$
$$\ldots\left[-k_a\left(\frac{D_1}{D_2}\sqrt{\frac{\alpha_2}{\alpha_1}}+1\right)\right]\cos\left(\left[\frac{a}{\sqrt{\alpha_1}}+\frac{b-a}{\sqrt{\alpha_2}}\right]\beta_n\right)+\left[k_a\left(\frac{D_1}{D_2}\sqrt{\frac{\alpha_2}{\alpha_1}}-1\right)\right]\cos\left(\left[\frac{a}{\sqrt{\alpha_1}}-\frac{b-a}{\sqrt{\alpha_2}}\right]\beta_n\right)=0 \tag{5-98}$$

Thus, the eigenvalues for the boundary conditions of case 2 are given by the roots of the above equation.

5.4.4 Initial conditions

The complete solution to the system, $c_{A,i}(z,t)$ in each layer is constructed by a linear superposition of product of the temporal and spatial solutions.

$$\begin{aligned} c_{A,i}(z,t) &= \sum_{n=1}^{\infty}\delta_n\cdot\Psi_i(\beta_n,z)\cdot\Gamma(\beta_n,t) \\ &= \sum_{n=1}^{\infty}\delta_n\cdot\Psi_i(\beta_n,z)\exp\left[-\beta_n{}^2t\right] \end{aligned} \tag{5-99}$$

where δ_n are constant coefficients used to satisfy the initial boundary conditions. The values of δ_n can be explicitly determined by solving equation (5-99) at $t = 0$.

$$c_{A,i}(z,t)\Big|_{t=0} = \sum_{n=1}^{\infty}\delta_n\cdot\Psi_i(\beta_{n,}z) = c_{A0}(z) \tag{5-100}$$

and using the property that the eigenfunctions are orthogonal,

$$R_{f,1}\int_0^a \Psi_1(\beta_m,z)\cdot\Psi_1(\beta_n,z)dz + R_{f,2}\int_a^b \Psi_2(\beta_m,z)\cdot\Psi_2(\beta_n,z)dz = \begin{cases}0 & \text{for } m \neq n \\ N(\beta_n) & \text{for } m = n\end{cases} \tag{5-101}$$

where the *normalization integral*, $N(\beta_n)$, is defined as by the following expression:

$$N(\beta_n) = R_{f,1}\int_0^a [\Psi_1(\beta_n,z)]^2 dz + R_{f,2}\int_a^b [\Psi_2(\beta_n,z)]^2 dz \tag{5-102}$$

Operating on both sides of equation (5-100) with the operator $R_{f,1}\int_0^a \Psi_1(\beta_r,z)dz$ for $z \in [0,a]$ and $R_{f,2}\int_a^b \Psi_2(\beta_r,z)dz$ for $z \in [a,b]$ for an arbitrary r. Summing these we obtain

$$\begin{aligned}&\sum_{n=1}^{\infty}\delta_n\left[R_{f,1}\int_0^a \Psi_1(\beta_r,z)\cdot\Psi_1(\beta_{n,}z)dz + R_{f,2}\int_a^b \Psi_2(\beta_r,z)\cdot\Psi_2(\beta_{n,}z)dz\right] \\ &= R_{f,1}\int_0^a \Psi_1(\beta_r,z)\cdot c_{A0}(z)dz + R_{f,2}\int_a^b \Psi_2(\beta_r,z)\cdot c_{A0}(z)dz\end{aligned} \tag{5-103}$$

The right-hand side of equation (5-103) is the *initialization integral*, $I_0(\beta_n)$, as defined by equation (5-9). The left-hand side can be simplified using the orthogonality property given in equation (5-101). Thus we have nonzero values only when $r = n$, hence

$$\delta_n \cdot N(\beta_n) = I_0(\beta_n) \tag{5-104}$$

Substituting for the constant coefficient, the generalized solution for the concentration thus is given by

$$c_{A,i}(z,t) = \sum_{n=1}^{\infty}\frac{1}{N(\beta_n)}\cdot I_0(\beta_n)\cdot\Psi_i(\beta_n,z)\exp\left[-\beta_n{}^2 t\right] \tag{5-105}$$

Differentiating equation (5-105) at $z = b$ gives a general solution for the surface flux

$$\begin{aligned} j_A(t)\big|_{z=b} &= -D_{A(eff),2}\left.\frac{\partial c_A}{\partial z}\right|_{z=b} \\ &= -D_{A(eff),2}\sum_{n=1}^{\infty}\left\{\frac{1}{N(\beta_n)}\cdot I_0(\beta_n)\cdot\left.\frac{\partial\Psi_2}{\partial z}\right|_{z=b}\exp\left[-\beta_n{}^2 t\right]\right\}\end{aligned} \tag{5-106}$$

5.4.5 Variable transformation for first-order decay

Solving the system with the inclusion of a first-order decay of the contaminant species, as described by the dynamic equations (5-10) and (5-11), we use the following variable substitution:

$$c_{A,i}(z,t) = \hat{c}_i(z,t)\exp(-k_1 t) \tag{5-107}$$

Effect of transformation on the dynamic equation

The transformed left-hand side of equations (5-10) and (5-11), the partial derivative with respect to time, gives

$$\frac{\partial c_{A,i}}{\partial t} = -k_1\hat{c}_i(z,t)\exp(-k_1 t) + \frac{\partial \hat{c}_i}{\partial t}\exp(-k_1 t) \tag{5-108}$$

The transformed right-hand side, spatial terms, gives

$$\left(\frac{D_{A(eff),i}}{R_{f,i}}\right)\frac{\partial^2 c_{A,i}}{\partial z^2} - k_1 c_{A,i}(z,t) = \left(\frac{D_{A(eff),i}}{R_{f,i}}\right)\frac{\partial^2 \hat{c}_i}{\partial z^2}\exp(-k_1 t) - k_1\hat{c}_i(z,t)\exp(-k_1 t) \tag{5-109}$$

Hence, the transformed dynamic equations result in

$$\therefore \quad \frac{\partial \hat{c}_i}{\partial t} = \left(\frac{D_{A(eff),i}}{R_{f,i}}\right)\frac{\partial^2 \hat{c}_i}{\partial z^2} \tag{5-110}$$

The net effect of the transformation is to remove the reaction term from the dynamic equation, resulting in the standard diffusion-adsorption formulation.

Effect of transformation on initial conditions

At time equal to zero the transformed initial concentration results in

$$c_{A,i}(z,0) = \hat{c}_i(z,0)\exp(0) \equiv \hat{c}_i(z,0) \tag{5-111}$$

Hence, all initial conditions remain unaltered after the transformation.

Effect of the transformation on boundary conditions

Transformation of the boundary conditions at the interface and the bottom boundary results in

$$\hat{c}_1(z,t)\exp(-k_1 t) = \hat{c}_2(z,t)\exp(-k_1 t) \qquad \text{at } z = a,\ t > 0 \tag{5-112}$$

$$D_{A(eff),1}\frac{\partial \hat{c}_1}{\partial z}\exp(-k_1 t) = D_{A(eff),2}\frac{\partial \hat{c}_2}{\partial z}\exp(-k_1 t) \qquad \text{at } z = a,\ t > 0 \tag{5-113}$$

$$\frac{\partial \hat{c}_1}{\partial z}\exp(-k_1 t) = 0 \qquad \text{at } z = 0,\ t > 0 \tag{5-114}$$

Transformation of the upper boundary condition for the case of zero surface concentration gives

$$\hat{c}_2(z,t)\exp(-k_1 t) = 0 \qquad \text{at } z = b,\ t > 0 \tag{5-115}$$

and for a mass transfer upper boundary gives

$$D_{A(eff),2}\frac{\partial \hat{c}_2}{\partial z}\exp(-k_1 t)+k_a\cdot \hat{c}_2(z,t)\exp(-k_1 t)=0 \qquad \text{at } z=b,\ t>0 \quad (5\text{-}116)$$

Dividing all these equations by the common factor $\exp(-k_1 t)$ shows that the boundary conditions remain unaltered after the transformation.

Effect on the solution and surface flux

As the transformed system of equations results in the standard formulation with diffusion and adsorption mechanisms only, we solve for $\hat{c}_i(z,t)$ as described in the previous sections of 5.4. The resulting solution then is inverted back into terms of $c_{A,i}(z,t)$ by equation (5-107).

The net effect of the decay term is to include an additional exponential time term to the solution. Since the flux is determined by a differential with respect to space, it too is only modified by this additional factor.

References

Özisk, M.N. (1993) *Heat Conduction*, 2nd ed., John Wiley & Sons, New York.

Carslaw, H.S., Jaeger, J.C. (1959) *Conduction of Heat in Solids*, Clarendon Press, Oxford.

6 Diffusion in a three-layer composite system

6.1 Introduction

Solutions to the diffusion equations in a geometry of three or more layers are rarely available. These problems may arise in layered soil systems. Multilayered composites may also be used as capping material over contaminated sediment (Thoma et al., 1997). In these caps, a highly adsorbing material (e.g., high organic carbon content sediments) is placed near the contaminated zone, while sandy material covers over the top. The thin layer of absorbing material effectively retards the contaminant migration, while the sand layer provides both stabilization of this absorbing material and protection from surface bioturbation. A multilayer composite model may also be used to model contaminant vapor transport in a sediment column with a low moisture content profile. Different layers can represent various moisture content regimes, which consequently have different contaminant transport properties.

The solutions giving the mobile phase concentration profile and surface flux rate for a system consisting of a composite of three layers of different properties is presented. Cases where surface mass transfer and contaminant decay are significant and insignificant are both analyzed. These solutions are given in Section 6.2. A brief discussion on the numerical evaluation of the solution is given in Section 6.3. The system equations were solved using a separation of variables technique. The solution method is outlined in Section 6.4.

6.2 Analysis summary

6.2.1 System dynamics and general solution for a three-layer composite

The dynamics of a contaminant species A due to diffusion and adsorption processes, in a finite soil/sediment layer $z \in [0, z_3]$ which contains three distinct zones, is defined by the following equations:

$$\frac{\partial c_{A,1}}{\partial t} = \left(\frac{D_{A(eff),1}}{R_{f,1}}\right)\frac{\partial^2 c_{A,1}}{\partial z^2} \qquad z \in [0, z_1] \qquad (6\text{-}1)$$

$$\frac{\partial c_{A,2}}{\partial t} = \left(\frac{D_{A(eff),2}}{R_{f,2}}\right)\frac{\partial^2 c_{A,2}}{\partial z^2} \qquad z \in [z_1, z_2] \qquad (6\text{-}2)$$

$$\frac{\partial c_{A,3}}{\partial t} = \left(\frac{D_{A(eff),3}}{R_{f,3}}\right)\frac{\partial^2 c_{A,3}}{\partial z^2} \qquad z \in [z_2, z_3] \qquad (6\text{-}3)$$

where $c_A(z,t)$ is the concentration of species A in the mobile pore air or pore water phase and $z = \{z_1, z_2\}$ are the locations of the interfaces between the three layers. Perfect contacting is assumed at the interface between the separate layers, hence the concentration at that position is the identical and no accumulation occurs. A no-flow boundary condition is assumed the bottom of the system $z = 0$.

Mathematically these boundary conditions are expressed as

$$\frac{\partial c_{A,1}}{\partial z}=0 \qquad \text{at } z=0,\ t>0 \tag{6-4}$$

$$D_{A(eff),1}\frac{\partial c_{A,1}}{\partial z}=D_{A(eff),2}\frac{\partial c_{A,2}}{\partial z} \qquad \text{at } z=z_1,\ t>0 \tag{6-5}$$

$$c_{A,1}(z,t)=c_{A,2}(z,t) \qquad \text{at } z=z_1,\ t>0 \tag{6-6}$$

$$D_{A(eff),2}\frac{\partial c_{A,2}}{\partial z}=D_{A(eff),3}\frac{\partial c_{A,3}}{\partial z} \qquad \text{at } z=z_2,\ t>0 \tag{6-7}$$

$$c_{A,2}(z,t)=c_{A,3}(z,t) \qquad \text{at } z=z_2,\ t>0 \tag{6-8}$$

Arbitrary initial conditions are specified by the following:

$$c_{A,1}(z,t)=c_{A0}(z) \qquad \text{at } t=0,\ z\in[0,z_1] \tag{6-9}$$

$$c_{A,2}(z,t)=c_{A0}(z) \qquad \text{at } t=0,\ z\in[z_1,z_2] \tag{6-10}$$

$$c_{A,3}(z,t)=c_{A0}(z) \qquad \text{at } t=0,\ z\in[z_2,z_3] \tag{6-11}$$

The general solution for the mobile phase concentration with the dynamics determined by equations (6-1) to (6-3) is given by

$$c_{A,i}(z,t)=\sum_{n=1}^{\infty}\left\{\exp\left(-\beta_n^{\,2}t\right)\cdot\frac{1}{N(\beta_n)}\cdot\Psi_i(\beta_n,z)\cdot I_0(\beta_n)\right\} \tag{6-12}$$

and surface flux from the top surface, $z=z_3$, is given by

$$j_A(t)\Big|_{z=z_3}=-D_{A(eff),3}\sum_{n=1}^{\infty}\left\{\exp\left(-\beta_n^{\,2}t\right)\cdot\frac{1}{N(\beta_n)}\cdot\frac{\partial\Psi_3}{\partial z}\bigg|_{z=z_3}\cdot I_0(\beta_n)\right\} \tag{6-13}$$

where $\Psi_i(\beta_n,z)$ and β_n are the *eigenfunctions* and *eigenvalues* which depend on which boundary condition is used. The generalized form of the eigenfunction is described in Section 6.2.2. The specific formulation is given for the case of zero surface concentration at the boundary in Section 6.2.3, and for a mass transfer (or reaction) boundary condition at the surface in Section 6.2.4.

For the above general solution, $N(\beta_n)$ is the *normalization integral* given by

$$N(\beta_n)=R_{f,1}\int_0^{z_1}\left[\Psi_1(\beta_n,z)\right]^2dz+R_{f,2}\int_{z_1}^{z_2}\left[\Psi_2(\beta_n,z)\right]^2dz+R_{f,3}\int_{z_2}^{z_3}\left[\Psi_3(\beta_n,z)\right]^2dz \tag{6-14}$$

and $I_0(\beta_n)$ is the *initialization integral* given by

$$I_0(\beta_n) = R_{f,1}\int_0^{z_1}\Psi_1(\beta_n,z)\cdot c_{A0}(z)dz + R_{f,2}\int_{z_1}^{z_2}\Psi_2(\beta_n,z)\cdot c_{A0}(z)dz + R_{f,3}\int_{z_2}^{z_3}\Psi_3(\beta_n,z)\cdot c_{A0}(z)dz \tag{6-15}$$

where $c_{A0}(z)$ are the initial conditions.

If a first-order decay mechanism of the contaminant is also of significance, the governing dynamic equations (6-1) to (6-3) are replaced by the following:

$$\frac{\partial c_{A,1}}{\partial t} = \left(\frac{D_{A(eff),1}}{R_{f,1}}\right)\frac{\partial^2 c_{A,1}}{\partial z^2} - k_1 c_{A,1}(z,t) \qquad z \in [0, z_1] \tag{6-16}$$

$$\frac{\partial c_{A,2}}{\partial t} = \left(\frac{D_{A(eff),2}}{R_{f,2}}\right)\frac{\partial^2 c_{A,2}}{\partial z^2} - k_1 c_{A,2}(z,t) \qquad z \in [z_1, z_2] \tag{6-17}$$

$$\frac{\partial c_{A,3}}{\partial t} = \left(\frac{D_{A(eff),3}}{R_{f,3}}\right)\frac{\partial^2 c_{A,3}}{\partial z^2} - k_1 c_{A,3}(z,t) \qquad z \in [z_2, z_3] \tag{6-18}$$

The general solution for the mobile phase concentration with the dynamics determined by equations (6-16) to (6-18) is given by

$$c_{A,i}(z,t) = \exp(-k_1 t)\sum_{n=1}^{\infty}\left\{\exp\left(-\beta_n^{\,2}t\right)\cdot\frac{1}{N(\beta_n)}\cdot\Psi_i(\beta_n,z)\cdot I_0(\beta_n)\right\} \tag{6-19}$$

and surface flux from the top surface, $z = z_3$, is given by

$$j_A(t)\big|_{z=z_3} = -D_{A(eff),3}\exp(-k_1 t)\sum_{n=1}^{\infty}\left\{\exp\left(-\beta_n^{\,2}t\right)\cdot\frac{1}{N(\beta_n)}\cdot\left.\frac{\partial\Psi_3}{\partial z}\right|_{z=z_3}\cdot I_0(\beta_n)\right\} \tag{6-20}$$

where $\Psi_i(\beta_n,z)$ and β_n are the eigenfunctions and eigenvalues of the system and the normalization and initialization integral is defined in equations (6-14) and (6-15) above. The eigenfunctions and eigenvalues are defined in the same manner as that for the cases without the decay process dynamics.

6.2.2 System eigenfunctions and eigenvalues

The eigenfunctions are an orthogonal set of functions used to generate the overall concentration profile. The one-dimensional composite diffusion eigenfunctions have the general form of

$$\Psi_i(\beta_n,z) = A_{i,n}\sin\left(\sqrt{\frac{R_{f,i}}{D_{A(eff),i}}}\cdot\beta_n\cdot z\right) + B_{i,n}\cos\left(\sqrt{\frac{R_{f,i}}{D_{A(eff),i}}}\cdot\beta_n\cdot z\right) \quad (6\text{-}21)$$

where the eigenvalues, $\beta_n : n = 1,2,3\ldots$, and eigenfunction coefficients, $A_{i,n}$ and $B_{i,n}$, are determined by the system boundary conditions.

The following are forms of the eigenfunction that are required for the calculation for the normalization and initialization integrals, and hence the concentration and surface flux. Define

$$\lambda_{i,n} = \beta_n\sqrt{\frac{R_{f,i}}{D_{A(eff),i}}} \quad (6\text{-}22)$$

The partial differential of the generalized eigenfunction with respect to z is

$$\frac{\partial\Psi_i}{\partial z} = \{\lambda_{i,n}A_{i,n}\}\cos(\lambda_{i,n}z) + \{-\lambda_{i,n}B_{i,n}\}\sin(\lambda_{i,n}z) \quad (6\text{-}23)$$

The indefinite integral of the generalized eigenfunction with respect to z is

$$\int\Psi(\beta_n,z)dz = \left\{-\frac{A_{i,n}}{\lambda_{i,n}}\right\}\cos(\lambda_{i,n}z) + \left\{\frac{B_{i,n}}{\lambda_{i,n}}\right\}\sin(\lambda_{i,n}z) \quad (6\text{-}24)$$

The square of the generalized eigenfunction is

$$[\Psi_i(\beta_n,z)]^2 = A_{i,n}{}^2\sin^2(\lambda_{i,n}z) + 2A_{i,n}B_{i,n}\sin(\lambda_{i,n}z)\cos(\lambda_{i,n}z) + B_{i,n}\cos^2(\lambda_{i,n}z) \quad (6\text{-}25)$$

The integral of the square of the generalized eigenfunction is

$$\int[\Psi_i(\beta_n,z)]^2 dz = \frac{\left[A_{i,n}{}^2 + B_{i,n}{}^2\right]\lambda_{i,n}z + \left[B_{i,n}{}^2 - A_{i,n}{}^2\right]\cos(\lambda_{i,n}z)\sin(\lambda_{i,n}z) + 2A_{i,n}B_{i,n}\sin^2(\lambda_{i,n}z)}{2\lambda_{i,n}} \quad (6\text{-}26)$$

The normalization integral may be evaluated by use of the expression given in equation (6-26). This is further discussed in Section 6.3.6. The initialization integral may be evaluated by use of the expression given in equation (6-24). This is further discussed in Section 6.3.7. The surface flux requires the evaluation of equation (6-23).

6.2.3 Case 1: Three-layer finite system with arbitrary initial concentrations, zero concentration at the surface, and zero flux at the base

A system is defined with the surface boundary condition

$$c_{A,3}(z,t) = 0 \qquad \text{at } z = z_3,\ t > 0 \qquad (6\text{-}27)$$

Other boundary conditions and initial conditions are given by equations (6-4) to (6-11).

With the dynamics given by equations (6-1) to (6-3), the system is illustrated by the following diagram:

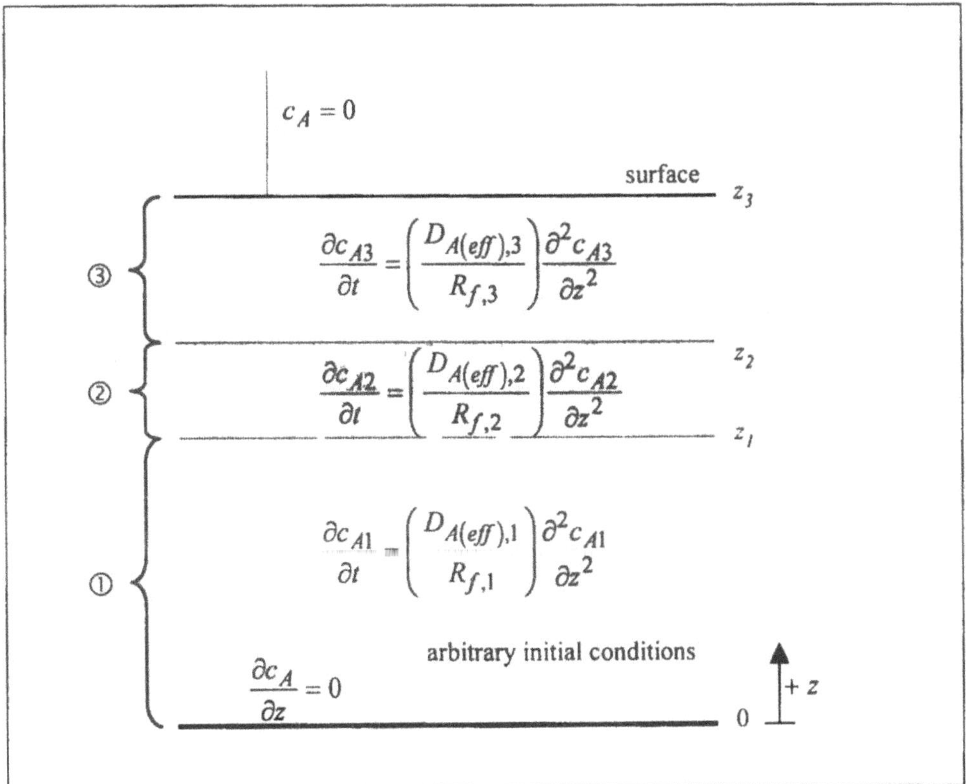

Figure 6-1 Diffusion in a three-layer finite system with arbitrary initial concentrations, zero concentration at the surface, and zero flux at the base.

With the dynamics given by equations (6-16) to (6-18), the system is illustrated by the following diagram:

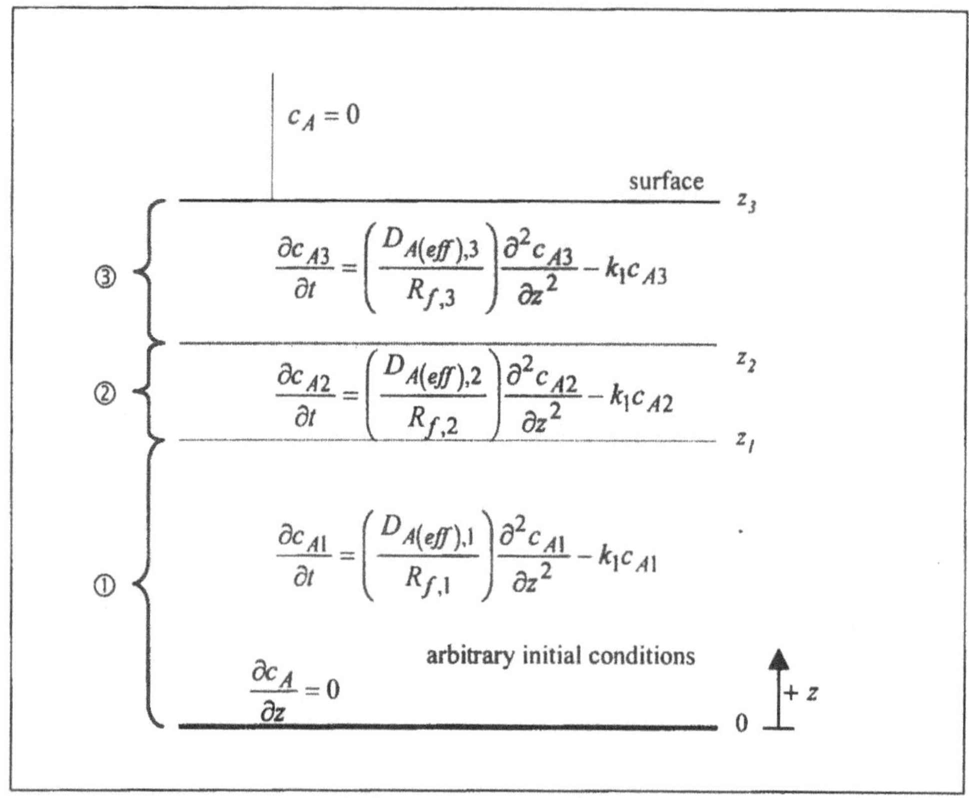

Figure 6-2 Diffusion in a three-layer finite system with arbitrary initial concentrations, zero concentration at the surface, zero flux at the base, and first-order decay.

In both dynamic situations the coefficients for the eigenfunctions are given by

$$A_{1,n} = 0 \tag{6-28}$$

$$B_{1,n} = 1 \tag{6-29}$$

$$A_{2,n} = \frac{-\cos\left(\frac{z_1}{\sqrt{\alpha_1}}\beta_n\right)\left\{[\gamma_3+\gamma_2]\cos\left(\beta_n\left[\frac{z_2}{\sqrt{\alpha_2}}+\frac{z_3-z_2}{\sqrt{\alpha_3}}\right]\right)+[\gamma_3-\gamma_2]\cos\left(\beta_n\left[\frac{z_2}{\sqrt{\alpha_2}}-\frac{z_3-z_2}{\sqrt{\alpha_3}}\right]\right)\right\}}{[\gamma_3+\gamma_2]\sin\left(\beta_n\left[\frac{z_3-z_2}{\sqrt{\alpha_3}}+\frac{z_2-z_1}{\sqrt{\alpha_2}}\right]\right)-[\gamma_3-\gamma_2]\sin\left(\beta_n\left[\frac{z_3-z_2}{\sqrt{\alpha_3}}-\frac{z_2-z_1}{\sqrt{\alpha_2}}\right]\right)} \tag{6-30}$$

$$B_{2,n}=\frac{\cos\left(\frac{z_1}{\sqrt{\alpha_1}}\beta_n\right)\left\{[\gamma_3+\gamma_2]\sin\left(\beta_n\left[\frac{z_2}{\sqrt{\alpha_2}}+\frac{z_3-z_2}{\sqrt{\alpha_3}}\right]\right)+[\gamma_3-\gamma_2]\sin\left(\beta_n\left[\frac{z_2}{\sqrt{\alpha_2}}-\frac{z_3-z_2}{\sqrt{\alpha_3}}\right]\right)\right\}}{[\gamma_3+\gamma_2]\sin\left(\beta_n\left[\frac{z_3-z_2}{\sqrt{\alpha_3}}+\frac{z_2-z_1}{\sqrt{\alpha_2}}\right]\right)-[\gamma_3-\gamma_2]\sin\left(\beta_n\left[\frac{z_3-z_2}{\sqrt{\alpha_3}}-\frac{z_2-z_1}{\sqrt{\alpha_2}}\right]\right)} \tag{6-31}$$

$$A_{3,n}=\frac{-2\gamma_2\cos\left(\frac{z_1}{\sqrt{\alpha_1}}\beta_n\right)\cos\left(\frac{z_3}{\sqrt{\alpha_3}}\beta_n\right)}{[\gamma_3+\gamma_2]\sin\left(\beta_n\left[\frac{z_3-z_2}{\sqrt{\alpha_3}}+\frac{z_2-z_1}{\sqrt{\alpha_2}}\right]\right)-[\gamma_3-\gamma_2]\sin\left(\beta_n\left[\frac{z_3-z_2}{\sqrt{\alpha_3}}-\frac{z_2-z_1}{\sqrt{\alpha_2}}\right]\right)} \tag{6-32}$$

$$B_{3,n}=\frac{2\gamma_2\cos\left(\frac{z_1}{\sqrt{\alpha_1}}\beta_n\right)\sin\left(\frac{z_3}{\sqrt{\alpha_3}}\beta_n\right)}{[\gamma_3+\gamma_2]\sin\left(\beta_n\left[\frac{z_3-z_2}{\sqrt{\alpha_3}}+\frac{z_2-z_1}{\sqrt{\alpha_2}}\right]\right)-[\gamma_3-\gamma_2]\sin\left(\beta_n\left[\frac{z_3-z_2}{\sqrt{\alpha_3}}-\frac{z_2-z_1}{\sqrt{\alpha_2}}\right]\right)} \tag{6-33}$$

where $\alpha_i=\left(\frac{D_{A(eff),i}}{R_{f,i}}\right)$ and $\gamma_i=\sqrt{D_{A(eff),i}\cdot R_{f,i}}$

The eigenvalues β_n: $n=1,2,3\ldots$ are the positive roots of

$$\begin{aligned}&\left[\gamma_1\gamma_2+\gamma_1\gamma_3+\gamma_2^2+\gamma_2\gamma_3\right]\cos\left(\beta_n\left[\frac{z_3-z_2}{\sqrt{\alpha_3}}+\frac{z_2-z_1}{\sqrt{\alpha_2}}+\frac{z_1}{\sqrt{\alpha_1}}\right]\right)\ldots\\&\ldots+\left[-\gamma_1\gamma_2-\gamma_1\gamma_3+\gamma_2^2+\gamma_2\gamma_3\right]\cos\left(\beta_n\left[\frac{z_3-z_2}{\sqrt{\alpha_3}}+\frac{z_2-z_1}{\sqrt{\alpha_2}}-\frac{z_1}{\sqrt{\alpha_1}}\right]\right)\ldots\\&\ldots+\left[\gamma_1\gamma_2-\gamma_1\gamma_3-\gamma_2^2+\gamma_2\gamma_3\right]\cos\left(\beta_n\left[\frac{z_3-z_2}{\sqrt{\alpha_3}}-\frac{z_2-z_1}{\sqrt{\alpha_2}}+\frac{z_1}{\sqrt{\alpha_1}}\right]\right)\ldots\\&\ldots+\left[-\gamma_1\gamma_2+\gamma_1\gamma_3-\gamma_2^2+\gamma_2\gamma_3\right]\cos\left(\beta_n\left[\frac{z_3-z_2}{\sqrt{\alpha_3}}-\frac{z_2-z_1}{\sqrt{\alpha_2}}-\frac{z_1}{\sqrt{\alpha_1}}\right]\right)=0\end{aligned} \tag{6-34}$$

These eigenvalues and eigenfunction coefficients are substituted into the system eigenfunction as given in Section 6.2.2 and are then used to evaluate the concentration or surface flux, as described in Section 6.2.1.

6.2.4 Case 2: Three-layer finite system with arbitrary initial concentrations, mass transfer or reaction at the surface, and zero flux at the base

A system is defined with the surface boundary condition

$$D_{A(eff),3}\frac{\partial c_{A,3}}{\partial z}+k_a\cdot c_A(z,t)=0 \qquad \text{at } z=z_3,\ t>0 \qquad (6\text{-}35)$$

Other boundary conditions and initial conditions are given by equations (6-4) to (6-11).

With the dynamics given by equations (6-1) to (6-3), the system is illustrated by the following diagram:

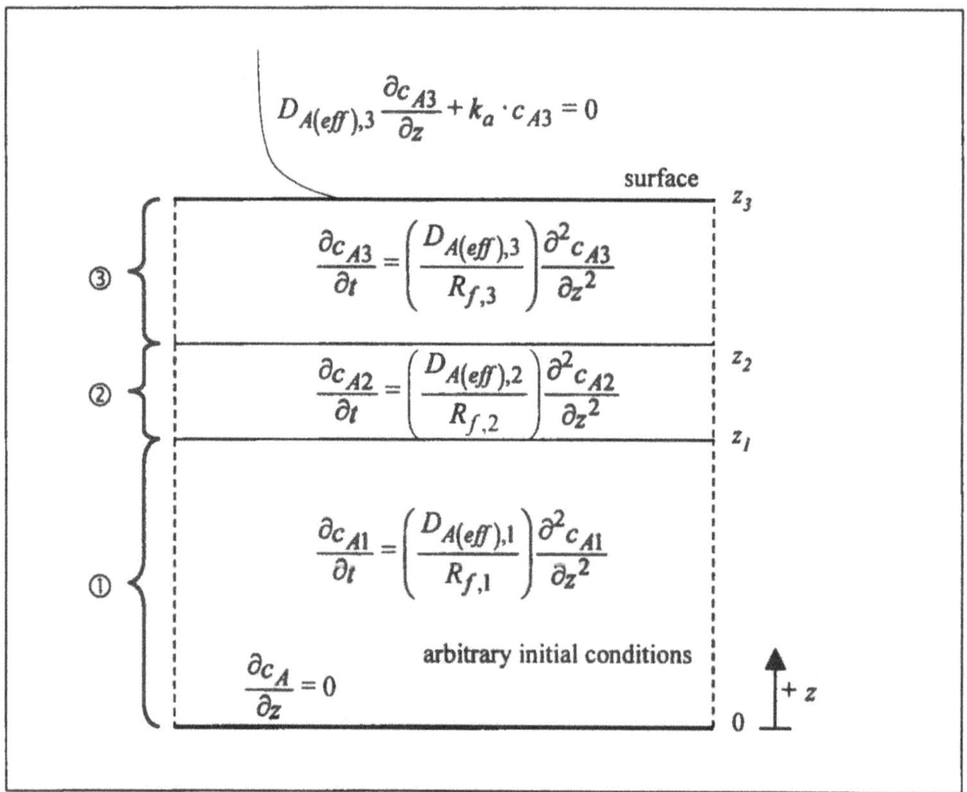

Figure 6-3 Diffusion in a three-layer finite system with arbitrary initial concentrations, mass transfer or reaction at the surface, and zero flux at the base.

With the dynamics given by equations (6-16) to (6-18), the system is illustrated by the following diagram:

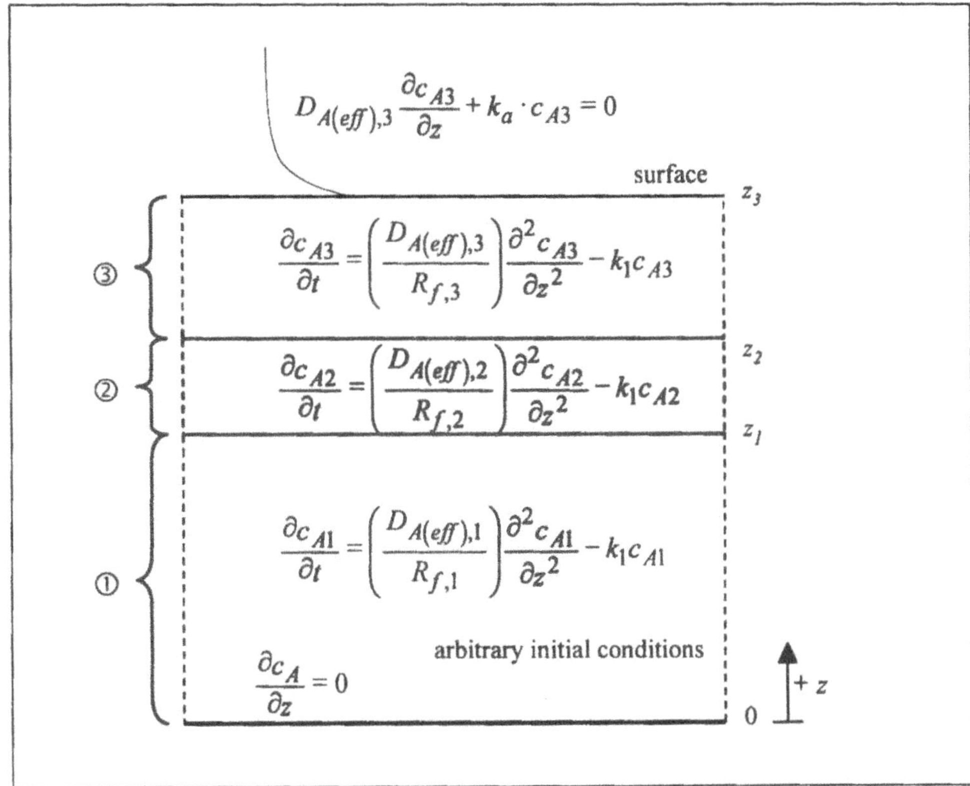

Figure 6-4 Diffusion in a three-layer finite system with arbitrary initial concentrations, mass transfer or reaction at the surface, zero flux at the base, and first-order decay.

In both dynamic situations the coefficients for the eigenfunctions are given by

$$A_{1,n}=0 \tag{6-36}$$

$$B_{1,n}=1 \tag{6-37}$$

$$A_{2,n}=\frac{1}{\Delta}\left\{\begin{array}{l}k_a\beta_n\cos\left(\frac{z_1}{\sqrt{\alpha_1}}\beta_n\right)\left\{\gamma_2\sin\left(\frac{z_2}{\sqrt{\alpha_2}}\beta_n\right)\sin\left(\frac{z_3-z_2}{\sqrt{\alpha_3}}\beta_n\right)-\gamma_3\cos\left(\frac{z_2}{\sqrt{\alpha_2}}\beta_n\right)\cos\left(\frac{z_3-z_2}{\sqrt{\alpha_3}}\beta_n\right)\right\}\\+\gamma_3\beta_n^2\cos\left(\frac{z_1}{\sqrt{\alpha_1}}\beta_n\right)\left\{\gamma_2\sin\left(\frac{z_2}{\sqrt{\alpha_2}}\beta_n\right)\cos\left(\frac{z_3-z_2}{\sqrt{\alpha_3}}\beta_n\right)+\gamma_3\cos\left(\frac{z_2}{\sqrt{\alpha_2}}\beta_n\right)\sin\left(\frac{z_3-z_2}{\sqrt{\alpha_3}}\beta_n\right)\right\}\end{array}\right\} \tag{6-38}$$

$$B_{2,n}=\frac{1}{\Delta}\left\{\begin{aligned}&k_a\beta_n\cos\left(\frac{z_1}{\sqrt{\alpha_1}}\beta_n\right)\left\{\gamma_2\cos\left(\frac{z_2}{\sqrt{\alpha_2}}\beta_n\right)\sin\left(\frac{z_3-z_2}{\sqrt{\alpha_3}}\beta_n\right)+\gamma_3\sin\left(\frac{z_2}{\sqrt{\alpha_2}}\beta_n\right)\cos\left(\frac{z_3-z_2}{\sqrt{\alpha_3}}\beta_n\right)\right\}\\&+\gamma_3\beta_n^2\cos\left(\frac{z_1}{\sqrt{\alpha_1}}\beta_n\right)\left\{\gamma_2\cos\left(\frac{z_2}{\sqrt{\alpha_2}}\beta_n\right)\cos\left(\frac{z_3-z_2}{\sqrt{\alpha_3}}\beta_n\right)-\gamma_3\sin\left(\frac{z_2}{\sqrt{\alpha_2}}\beta_n\right)\sin\left(\frac{z_3-z_2}{\sqrt{\alpha_3}}\beta_n\right)\right\}\end{aligned}\right\}$$

(6-39)

$$A_{3,n}=\frac{1}{\Delta}\left\{-k_a\gamma_2\beta_n\cos\left(\frac{z_1}{\sqrt{\alpha_1}}\beta_n\right)\cos\left(\frac{z_3}{\sqrt{\alpha_3}}\beta_n\right)+\gamma_2\gamma_3\beta_n^2\cos\left(\frac{z_1}{\sqrt{\alpha_1}}\beta_n\right)\sin\left(\frac{z_3}{\sqrt{\alpha_3}}\beta_n\right)\right\}$$

(6-40)

$$B_{3,n}=\frac{1}{\Delta}\left\{k_a\gamma_2\beta_n\cos\left(\frac{z_1}{\sqrt{\alpha_1}}\beta_n\right)\sin\left(\frac{z_3}{\sqrt{\alpha_3}}\beta_n\right)+\gamma_2\gamma_3\beta_n^2\cos\left(\frac{z_1}{\sqrt{\alpha_1}}\beta_n\right)\cos\left(\frac{z_3}{\sqrt{\alpha_3}}\beta_n\right)\right\} \quad (6\text{-}41)$$

where $\alpha_i=\left(\frac{D_{A(eff),i}}{R_{f,i}}\right)$, $\gamma_i=\sqrt{D_{A(eff),i}\cdot R_{f,i}}$ and

$$\begin{aligned}\Delta=\gamma_3\beta_n^2&\left[\gamma_2\cos\left(\frac{z_3-z_2}{\sqrt{\alpha_3}}\beta_n\right)\cos\left(\frac{z_2-z_1}{\sqrt{\alpha_2}}\beta_n\right)-\gamma_3\sin\left(\frac{z_3-z_2}{\sqrt{\alpha_3}}\beta_n\right)\sin\left(\frac{z_2-z_1}{\sqrt{\alpha_2}}\beta_n\right)\right]\\+k_a\beta_n&\left[\gamma_2\sin\left(\frac{z_3-z_2}{\sqrt{\alpha_3}}\beta_n\right)\cos\left(\frac{z_2-z_1}{\sqrt{\alpha_2}}\beta_n\right)+\gamma_3\cos\left(\frac{z_3-z_2}{\sqrt{\alpha_3}}\beta_n\right)\sin\left(\frac{z_2-z_1}{\sqrt{\alpha_2}}\beta_n\right)\right]\end{aligned}$$

The eigenvalues, $\beta_n: n=1,2,3\ldots$, are the positive roots of

$$0=\quad k_a\left\{\begin{aligned}&\left[\gamma_1\gamma_2+\gamma_1\gamma_3+\gamma_2^2+\gamma_2\gamma_3\right]\cos\left(\beta_n\left[\frac{z_3-z_2}{\sqrt{\alpha_3}}+\frac{z_2-z_1}{\sqrt{\alpha_2}}+\frac{z_1}{\sqrt{\alpha_1}}\right]\right)\\&+\left[-\gamma_1\gamma_2-\gamma_1\gamma_3+\gamma_2^2+\gamma_2\gamma_3\right]\cos\left(\beta_n\left[\frac{z_3-z_2}{\sqrt{\alpha_3}}+\frac{z_2-z_1}{\sqrt{\alpha_2}}-\frac{z_1}{\sqrt{\alpha_1}}\right]\right)\\&+\left[\gamma_1\gamma_2-\gamma_1\gamma_3-\gamma_2^2+\gamma_2\gamma_3\right]\cos\left(\beta_n\left[\frac{z_3-z_2}{\sqrt{\alpha_3}}-\frac{z_2-z_1}{\sqrt{\alpha_2}}+\frac{z_1}{\sqrt{\alpha_1}}\right]\right)\\&+\left[-\gamma_1\gamma_2+\gamma_1\gamma_3-\gamma_2^2+\gamma_2\gamma_3\right]\cos\left(\beta_n\left[\frac{z_3-z_2}{\sqrt{\alpha_3}}-\frac{z_2-z_1}{\sqrt{\alpha_2}}-\frac{z_1}{\sqrt{\alpha_1}}\right]\right)\end{aligned}\right\}\ldots$$

$$
\ldots+\gamma_3\beta_n\left\{\begin{aligned}
&\left[-\gamma_1\gamma_2-\gamma_1\gamma_3-\gamma_2^2-\gamma_2\gamma_3\right]\sin\left(\beta_n\left[\frac{z_3-z_2}{\sqrt{\alpha_3}}+\frac{z_2-z_1}{\sqrt{\alpha_2}}+\frac{z_1}{\sqrt{\alpha_1}}\right]\right)\\
&+\left[\gamma_1\gamma_2+\gamma_1\gamma_3-\gamma_2^2-\gamma_2\gamma_3\right]\sin\left(\beta_n\left[\frac{z_3-z_2}{\sqrt{\alpha_3}}+\frac{z_2-z_1}{\sqrt{\alpha_2}}-\frac{z_1}{\sqrt{\alpha_1}}\right]\right)\\
&+\left[-\gamma_1\gamma_2+\gamma_1\gamma_3+\gamma_2^2-\gamma_2\gamma_3\right]\sin\left(\beta_n\left[\frac{z_3-z_2}{\sqrt{\alpha_3}}-\frac{z_2-z_1}{\sqrt{\alpha_2}}+\frac{z_1}{\sqrt{\alpha_1}}\right]\right)\\
&+\left[\gamma_1\gamma_2-\gamma_1\gamma_3+\gamma_2^2-\gamma_2\gamma_3\right]\sin\left(\beta_n\left[\frac{z_3-z_2}{\sqrt{\alpha_3}}-\frac{z_2-z_1}{\sqrt{\alpha_2}}-\frac{z_1}{\sqrt{\alpha_1}}\right]\right)
\end{aligned}\right\} \quad (6\text{-}42)
$$

These eigenvalues and eigenfunction coefficients are substituted into the system eigenfunction as given in Section 6.2.2 and are then used to evaluate the concentration or surface flux, as described in Section 6.2.1.

6.3 Numerical Evaluation

Both calculations for the concentration and surface flux require the sum of an infinite series of terms. As the terms tend to infinity, their effect on the solution decreases in significance. Numerically, the range in which the terms are still of significance needs to be determined. Following this, the eigenvalues in this range are evaluated. The eigenvalues are determined by finding roots to the appropriate transcendental functions determined by the system boundary conditions.

Once the eigenvalues have been determined, the eigenfunctions and hence the normalization integral, initialization integral, differential of the eigenfunction may be calculated. These are combined according to the solution given in equations (6-12) and (6-13), of if the contaminant decay dynamics is also included, equations (6-19) and (6-20).

6.3.1 Concentration calculation

The following diagram shows the stages needed for the calculation of a mobile phase concentration at position $z \in [0, z_3]$ and time $t > 0$. Each box in the diagram below represents a stage in the calculation process. It has references to the appropriate equations or section in this manuscript.

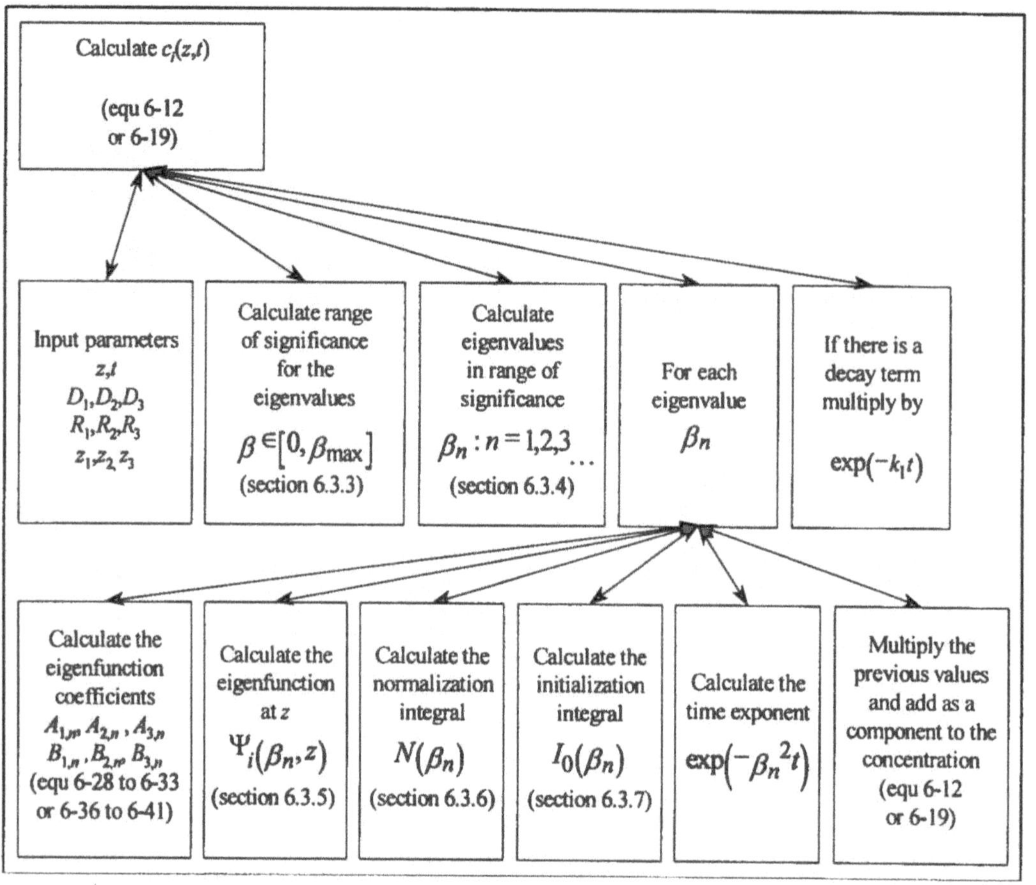

Figure 6-5 Information flow diagram for concentration calculation.

6.3.2 Surface flux calculation

The following diagram shows the stages needed for the calculation of the flux at the sediment surface at time $t > 0$. Each box in the diagram below represents a stage in the calculation process. It has references to the appropriate equations or section in this manuscript.

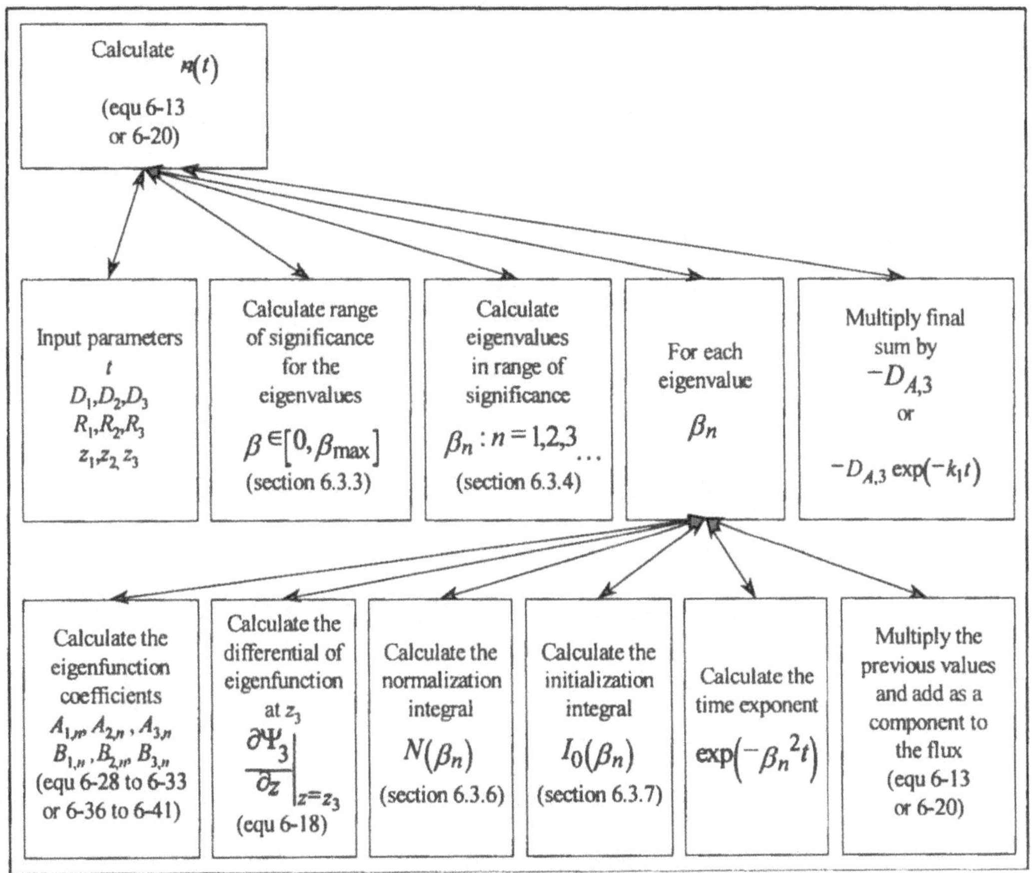

Figure 6-6 Information flow diagram for surface flux calculation.

6.3.3 Range of significance for eigenvalues

Both the concentration and flux equations require the calculation of an infinite number of eigenvalues, $\beta_n : n = 1,2,3...$ However as the negative square of the eigenvalue is present in the exponential term, the significance of each eigenvalue decreases with increasing n.

A range of significant eigenvalues, $\beta_n \in (0, \beta_{max}]$, needs to be determined in order to numerically evaluate the concentration and flux. Initially estimate β_{max} as equal to 20 to 30 times the period of the smallest coefficient in the transcendental function sine and cosine terms. That is

$$\beta_{max} = \omega \frac{\pi}{\min\{\text{'sin' or 'cos' coefficients}\}} \qquad \omega \sim 20-30 \qquad (6\text{-}43)$$

If more eigenvalues are needed to generate an acceptable solution increase the value of ω as appropriate.

6.3.4 Determination of eigenvalues in range

The eigenvalues are given for case 1 as the roots of the equation

$$C_1 \cos[C_2 \cdot \beta_n] + C_3 \cos[C_4 \cdot \beta_n] + C_5 \cos[C_6 \cdot \beta_n] + C_7 \cos[C_8 \cdot \beta_n] = 0 \quad (6\text{-}44)$$

or for case 2 as the roots of the equation

$$\begin{aligned} &C_1 \cos[C_2 \cdot \beta_n] + C_3 \cos[C_4 \cdot \beta_n] + C_5 \cos[C_6 \cdot \beta_n] + C_7 \cos[C_8 \cdot \beta_n] \\ &+\beta_n\{C_9 \cdot \sin[C_2 \cdot \beta_n] + C_{10} \sin[C_4 \cdot \beta_n] + C_{11} \sin[C_6 \cdot \beta_n] + C_{12} \sin[C_8 \cdot \beta_n]\} = 0 \end{aligned} \quad (6\text{-}45)$$

where the coefficients are given by

$$C_1 = \left[\gamma_1\gamma_2 + \gamma_1\gamma_3 + \gamma_2^2 + \gamma_2\gamma_3\right] \quad (6\text{-}46)$$

$$C_2 = \left[\frac{z_3 - z_2}{\sqrt{\alpha_3}} + \frac{z_2 - z_1}{\sqrt{\alpha_2}} + \frac{z_1}{\sqrt{\alpha_1}}\right] \quad (6\text{-}47)$$

$$C_3 = \left[-\gamma_1\gamma_2 - \gamma_1\gamma_3 + \gamma_2^2 + \gamma_2\gamma_3\right] \quad (6\text{-}48)$$

$$C_4 = \left[\frac{z_3 - z_2}{\sqrt{\alpha_3}} + \frac{z_2 - z_1}{\sqrt{\alpha_2}} - \frac{z_1}{\sqrt{\alpha_1}}\right] \quad (6\text{-}49)$$

$$C_5 = \left[\gamma_1\gamma_2 - \gamma_1\gamma_3 - \gamma_2^2 + \gamma_2\gamma_3\right] \quad (6\text{-}50)$$

$$C_6 = \left[\frac{z_3 - z_2}{\sqrt{\alpha_3}} - \frac{z_2 - z_1}{\sqrt{\alpha_2}} + \frac{z_1}{\sqrt{\alpha_1}}\right] \quad (6\text{-}51)$$

$$C_7 = \left[-\gamma_1\gamma_2 + \gamma_1\gamma_3 - \gamma_2^2 + \gamma_2\gamma_3\right] \quad (6\text{-}52)$$

$$C_8 = \left[\frac{z_3 - z_2}{\sqrt{\alpha_3}} - \frac{z_2 - z_1}{\sqrt{\alpha_2}} - \frac{z_1}{\sqrt{\alpha_1}}\right] \quad (6\text{-}53)$$

$$C_9 = \frac{\gamma_3}{k_a}\left[-\gamma_1\gamma_2 - \gamma_1\gamma_3 - \gamma_2^2 - \gamma_2\gamma_3\right] \quad (6\text{-}54)$$

$$C_{10} = \frac{\gamma_3}{k_a}\left[\gamma_1\gamma_2 + \gamma_1\gamma_3 - \gamma_2^2 - \gamma_2\gamma_3\right] \quad (6\text{-}55)$$

$$C_{11} = \frac{\gamma_3}{k_a}\left[-\gamma_1\gamma_2 + \gamma_1\gamma_3 + \gamma_2^2 - \gamma_2\gamma_3\right] \quad (6\text{-}56)$$

$$C_{12} = \frac{\gamma_3}{k_a}\left[\gamma_1\gamma_2 - \gamma_1\gamma_3 + \gamma_2^2 - \gamma_2\gamma_3\right] \quad (6\text{-}57)$$

Roots of these equations are calculated within the range $\beta_n \in (0, \beta_{max}]$.

A rigorous bracketing procedure is required to capture all the roots within this range. Bracket boundaries are determined from the maxima and minima for each of the oscillatory functions. Each of these bracketed regions is checked to determine if a root is present. If a root does exist, a numerical search routine is used to determine its value. Hence the following four series are calculated:

$$\left\{0,\frac{\pi}{C_2},\frac{2\pi}{C_2},\frac{3\pi}{C_2},\ldots\right\}\leq\beta_{max},\left\{0,\frac{\pi}{C_4},\frac{2\pi}{C_4},\frac{3\pi}{C_4},\ldots\right\}\leq\beta_{max},$$

$$\left\{0,\frac{\pi}{C_6},\frac{2\pi}{C_6},\frac{3\pi}{C_6},\ldots\right\}\leq\beta_{max},\left\{0,\frac{\pi}{C_8},\frac{2\pi}{C_8},\frac{3\pi}{C_8},\ldots\right\}\leq\beta_{max} \tag{6-58}$$

for case 1. Case 2 has the following maxima and minima for the addition to the previous series for the extra transcendental terms

$$\left\{\frac{\pi}{2C_2},\frac{3\pi}{2C_2},\frac{5\pi}{2C_2},\ldots\right\}\leq\beta_{max},\left\{\frac{\pi}{2C_4},\frac{3\pi}{2C_4},\frac{5\pi}{2C_4},\ldots\right\}\leq\beta_{max},$$

$$\left\{\frac{\pi}{2C_6},\frac{3\pi}{2C_6},\frac{5\pi}{2C_6},\ldots\right\}\leq\beta_{max},\left\{\frac{\pi}{2C_8},\frac{3\pi}{2C_8},\frac{5\pi}{2C_8},\ldots\right\}\leq\beta_{max} \tag{6-59}$$

The series of maxima and minima are then combined and sorted into a single series given by

$$\{b_1,b_2,b_3,b_4\ldots\}\leq\beta_{max} \qquad \text{where } b_i<b_{i+1} \text{ and } b_1=0 \tag{6-60}$$

Any redundant points are eliminated from the combined series. This series gives the search brackets to be tested.

Defining the function

$$f(b)=C_1\cos[C_2\cdot b]+C_3\cdot\cos[C_4\cdot b]+C_5\cos[C_6\cdot b]+C_7\cdot\cos[C_8\cdot b] \tag{6-61}$$

For each bracket the following condition is tested

$$\text{if } f(b_i)\cdot f(b_{i+1})\leq 0 \text{ then root exists between } [b_i,b_{i+1}] \tag{6-62}$$

If the root exists on the lower bound, i.e., $f(b_i)=0$, the eigenvalue is recorded. We then proceed to the next bracket. If the root exists on the upper bound, i.e., $f(b_{i+1})=0$, the eigenvalue is not recorded. We then proceed to the next bracket where it will be picked up by the algorithm. If the root lies between the bracket bounds, a bisection algorithm is used to find the eigenvalue. This value is recorded.

This process is continued until all the bracketed regions in the range $(0,\beta_{max}]$ have been tested. The resulting series of recorded roots are the eigenvalues, β_n, in this range of significance.

Further details about the algorithms for bisection searches and for finding the roots of the stated transcendental problem can be found in Appendix C.

6.3.5 Eigenfunction evaluation

Given an eigenvalue, β_n, and system parameters, the coefficients for the all the eigenfunctions can be evaluated from equations (6-28) to (6-33) for case 1 and equations (6-36) to (6-41) for case 2.

Given a position, z, the layer in which it is located may be determined. The appropriate layer eigenfunction can then be evaluated from equation (6-21).

6.3.6 Normalization integral evaluation

Given an eigenvalue, β_n, and system parameters, the coefficients for the all the eigenfunctions can be evaluated from equations (6-28) to (6-33) for case 1, and from equations (6-36) to (6-41) for case 2.

The normalization integral can then be calculated from equations (6-14) and (6-26).

6.3.7 Initialization integral evaluation

The method of evaluating the initialization integral, given by equation (6-15), depends on how the initial concentration profile is given. If a functional form is available, $c_{A0} = f(z)$, the integral may be evaluated analytically. However, it is more common that the initial concentration is specified by a tabulation of discrete values.

If constant, discrete values are used, the initialization integral can be broken up into the sum of a series of definite integrals. For example, given the initial concentration distribution in the following diagram:

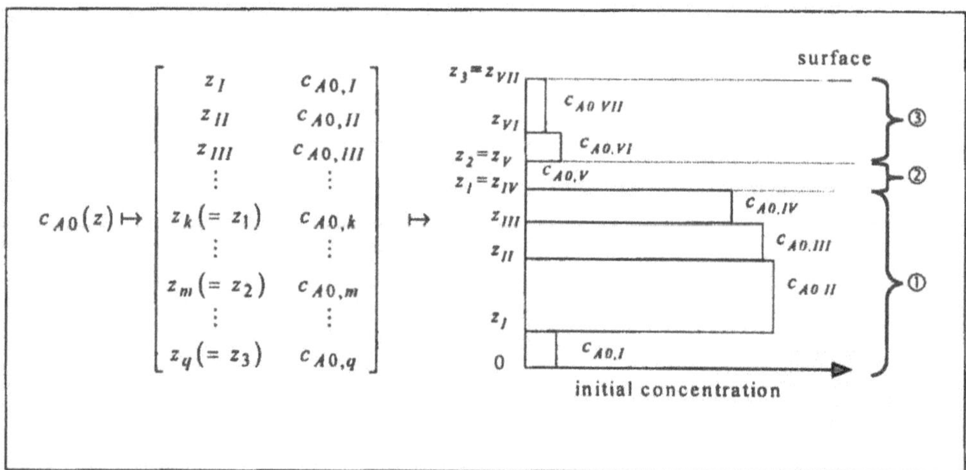

Figure 6-7 Method of specifying arbitrary initial conditions for a three-layer composite system.

The initialization integral can be calculated as

$$I_0(\beta_n) = R_{f,1}\left[\begin{array}{l} c_{A0,I}\int\limits_0^{z_I}\Psi_1(\beta_n,z)dz + c_{A0,II}\int\limits_{z_I}^{z_{II}}\Psi_1(\beta_n,z)dz + c_{A0,III}\int\limits_{z_{II}}^{z_{III}}\Psi_1(\beta_n,z)dz + \\ \ldots c_{A0,IV}\int\limits_{z_{III}}^{z_{IV}}\Psi_1(\beta_n,z)dz + c_{A0,V}\int\limits_{z_{IV}}^{z_V}\Psi_1(\beta_n,z)dz \end{array}\right]\ldots$$

$$\ldots + R_{f,2}\cdot c_{A0,VI}\int\limits_{z_V}^{z_{VI}}\Psi_2(\beta_n,z)dz \quad \ldots$$

$$\ldots + R_{f,3}\left[c_{A0,VII}\int\limits_{z_{VI}}^{z_{VII}}\Psi_3(\beta_n,z)dz + c_{A0,VIII}\int\limits_{z_{VII}}^{z_{VIII}}\Psi_3(\beta_n,z)dz\right] \tag{6-63}$$

where the definite integral of the eigenfunction is given by equation (6-24).

6.3.8 General numerical evaluation comments

The evaluation of the composite three-layer case is the most numerically intensive of the models in this manuscript. It involves numerically solving a difficult transcendental equation, which in turn is used to evaluate coefficients for further and further functions. As the calculation process continues, each stage in the numerical evaluation will increase the error of the results (due to numerical round off and truncation). The analyst should be careful to monitor for these problems which may give rise to nonphysical results such as oscillations in the solution or concentration profiles that seem to suggest negative diffusion coefficients. These problems are generaly worse when material properties or initial concentrations vary greatly.

If the system being modeled has three layers where the transport properties of adjacent layers are the same, this problem may degenerate to a two-layer or single-finite-layer model. All the finite-layer models can take arbitrary initial conditions. So, for example, a cap may simply be modeled by a single finite layer with the initial conditions set so as to have a zone of zero concentration at the top. This model will be both much simpler to evaluate and contain lower numerical uncertainty.

6.4 Development

6.4.1 Separation of variables

The system of partial differential equations is solved using a separation of variables technique. By the principles of this technique, the concentration is assumed to be separable into independent functions of position and time, of the form

$$c_{A,i}(z,t) = \Psi_i(z) \cdot \Gamma(t) \tag{6-64}$$

Substituting this into the dynamic equations (6-1), (6-2), and (6-3) we obtain

$$\alpha_i \frac{1}{\Psi_i(z)} \frac{d^2\Psi_i}{dz} = \frac{1}{\Gamma(t)} \frac{d\Gamma}{dt} \equiv -\beta^2 \tag{6-65}$$

where $\alpha_i = \left(\frac{D_{A(eff),i}}{R_{f,i}} \right)$. Note discussion on the effect of the decay term and the dynamic equations (6-16), (6-17), and (6-18) are deferred to Section 6.4.5.

As one side of equation (6-65) is a function of space only and the other a function of time only, these expressions must be equal to an arbitrary constant. Without loss of generality, this constant has been set as $-\beta^2$, where β is termed the separation factor. The spatial and temporal problems are now independently solved.

6.4.2 Solution to the temporal problem

The solution for the temporal equation given by

$$\frac{d\Gamma}{dt} = -\beta^2 \cdot \Gamma(t) \tag{6-66}$$

results in the following as a possible solution:

$$\Gamma(t) = \exp\left(-\beta^2 t\right) \tag{6-67}$$

6.4.3 Solution to the spatial problem

The solution for the spatial equations gives rise to eigenvalue problems, which depend on specific boundary conditions.

The eigenvalue problem can be stated as

$$\frac{d^2\Psi_{1,n}}{dz^2} + \frac{\beta_n^{\ 2}}{\alpha_1} \Psi_{1,n}(z) = 0 \qquad \text{for } z \in [0, z_1] \tag{6-68}$$

$$\frac{d^2\Psi_{2,n}}{dz^2} + \frac{\beta_n^{\ 2}}{\alpha_2} \Psi_{2,n}(z) = 0 \qquad \text{for } z \in [z_1, z_2] \tag{6-69}$$

$$\frac{d^2\Psi_{3,n}}{dz^2} + \frac{\beta_n^{\ 2}}{\alpha_3} \Psi_{3,n}(z) = 0 \qquad \text{for } z \in [z_2, z_3] \tag{6-70}$$

with transformed boundary conditions of

$$\frac{d\Psi_1}{dz} = 0 \qquad \text{at } z = 0 \tag{6-71}$$

$$D_1 \frac{d\Psi_1}{dz} = D_2 \frac{d\Psi_2}{dz} \qquad \text{at } z = z_l \tag{6-72}$$

$$\Psi_1(z) = \Psi_2(z) \qquad \text{at } z = z_1 \qquad (6\text{-}73)$$

$$D_2 \frac{d\Psi_2}{dz} = D_3 \frac{d\Psi_3}{dz} \qquad \text{at } z = z_2 \qquad (6\text{-}74)$$

$$\Psi_2(z) = \Psi_3(z) \qquad \text{at } z = z_2 \qquad (6\text{-}75)$$

with surface boundary condition given for case 1 by

$$\Psi_3(z) = 0 \qquad \text{at } z = z_3 \qquad (6\text{-}76)$$

and for case 2 by

$$D_3 \frac{d\Psi_3}{dz} + k_a \cdot \Psi_3(z) = 0 \qquad \text{at } z = z_3 \qquad (6\text{-}77)$$

Assuming the general form of

$$\Psi_i(\beta_n, z) = A_{i,n} \sin\left(\frac{\beta_n}{\sqrt{\alpha_i}} \cdot z\right) + B_{i,n} \cos\left(\frac{\beta_n}{\sqrt{\alpha_i}} \cdot z\right) \qquad (6\text{-}78)$$

hence

$$\frac{d\Psi_i}{dz} = \left\{\frac{\beta_n}{\sqrt{\alpha_i}} \cdot A_{i,n}\right\} \cos\left(\frac{\beta_n}{\sqrt{\alpha_i}} \cdot z\right) + \left\{-\frac{\beta_n}{\sqrt{\alpha_i}} \cdot B_{i,n}\right\} \sin\left(\frac{\beta_n}{\sqrt{\alpha_i}} \cdot z\right) \qquad (6\text{-}79)$$

The transformed boundary conditions are expressed in matrix form by the following:

$$\mathbf{A} \cdot \begin{bmatrix} A_{1,n} \\ B_{1,n} \\ A_{2,n} \\ B_{2,n} \\ A_{3,n} \\ B_{3,n} \end{bmatrix} = \begin{bmatrix} 0 \\ 0 \\ 0 \\ 0 \\ 0 \\ 0 \end{bmatrix} \qquad (6\text{-}80)$$

6.4.3.1 *Case 1 eigenvalue problem*

Matrix **A** for the transformed boundary conditions of equation (6-80) is given below.

With $\gamma_i = \sqrt{D_{A(eff),i} \cdot R_{f,i}}$

$$\mathbf{A} = \begin{bmatrix} \frac{\beta_n}{\sqrt{\alpha_1}} & 0 & 0 & 0 & 0 & 0 \\ \gamma_1 \beta_n \cos\left(\frac{z_1}{\sqrt{\alpha_1}}\beta_n\right) & -\gamma_1 \beta_n \sin\left(\frac{z_1}{\sqrt{\alpha_1}}\beta_n\right) & -\gamma_2 \beta_n \cos\left(\frac{z_1}{\sqrt{\alpha_2}}\beta_n\right) & \gamma_2 \beta_n \sin\left(\frac{z_1}{\sqrt{\alpha_2}}\beta_n\right) & 0 & 0 \\ \sin\left(\frac{z_1}{\sqrt{\alpha_1}}\beta_n\right) & \cos\left(\frac{z_1}{\sqrt{\alpha_1}}\beta_n\right) & -\sin\left(\frac{z_1}{\sqrt{\alpha_2}}\beta_n\right) & -\cos\left(\frac{z_1}{\sqrt{\alpha_2}}\beta_n\right) & 0 & 0 \\ 0 & 0 & \gamma_2 \beta_n \cos\left(\frac{z_2}{\sqrt{\alpha_2}}\beta_n\right) & -\gamma_2 \beta_n \sin\left(\frac{z_2}{\sqrt{\alpha_2}}\beta_n\right) & -\gamma_3 \beta_n \cos\left(\frac{z_2}{\sqrt{\alpha_3}}\beta_n\right) & \gamma_3 \beta_n \sin\left(\frac{z_2}{\sqrt{\alpha_3}}\beta_n\right) \\ 0 & 0 & \sin\left(\frac{z_2}{\sqrt{\alpha_2}}\beta_n\right) & \cos\left(\frac{z_2}{\sqrt{\alpha_2}}\beta_n\right) & -\sin\left(\frac{z_2}{\sqrt{\alpha_3}}\beta_n\right) & -\cos\left(\frac{z_2}{\sqrt{\alpha_3}}\beta_n\right) \\ 0 & 0 & 0 & 0 & \sin\left(\frac{z_3}{\sqrt{\alpha_3}}\beta_n\right) & \cos\left(\frac{z_3}{\sqrt{\alpha_3}}\beta_n\right) \end{bmatrix} \tag{6-81}$$

Solving for the coefficients using the transformed boundary conditions, it may be seen from the coefficients of **A** that $A_{1,n} = 0$. Thus $B_{1,n}$ may be set to any arbitrary constant. Let $B_{1,n} = 1$, thus the remaining coefficients are given by

$$\mathbf{B}\cdot\begin{bmatrix} A_{2,n} \\ B_{2,n} \\ A_{3,n} \\ B_{3,n} \end{bmatrix} = \begin{bmatrix} -\cos\left(\frac{z_1}{\sqrt{\alpha_1}}\beta_n\right) \\ 0 \\ 0 \\ 0 \end{bmatrix} \tag{6-82}$$

where the matrix **B** is given by

$$\mathbf{B} = \begin{bmatrix} -\sin\left(\frac{z_1}{\sqrt{\alpha_2}}\beta_n\right) & -\cos\left(\frac{z_1}{\sqrt{\alpha_2}}\beta_n\right) & 0 & 0 \\ \gamma_2\beta_n\cos\left(\frac{z_2}{\sqrt{\alpha_2}}\beta_n\right) & -\gamma_2\beta_n\sin\left(\frac{z_2}{\sqrt{\alpha_2}}\beta_n\right) & -\gamma_3\beta_n\cos\left(\frac{z_2}{\sqrt{\alpha_3}}\beta_n\right) & \gamma_3\beta_n\sin\left(\frac{z_2}{\sqrt{\alpha_3}}\beta_n\right) \\ \sin\left(\frac{z_2}{\sqrt{\alpha_2}}\beta_n\right) & \cos\left(\frac{z_2}{\sqrt{\alpha_2}}\beta_n\right) & -\sin\left(\frac{z_2}{\sqrt{\alpha_3}}\beta_n\right) & -\cos\left(\frac{z_2}{\sqrt{\alpha_3}}\beta_n\right) \\ 0 & 0 & \sin\left(\frac{z_3}{\sqrt{\alpha_3}}\beta_n\right) & \cos\left(\frac{z_3}{\sqrt{\alpha_3}}\beta_n\right) \end{bmatrix} \tag{6-83}$$

Solving for the coefficients using Cramer's rule, the determinant of the matrix **B** is

$$\begin{aligned} \Delta &= \gamma_2\beta_n\sin\left(\left[\frac{z_3-z_2}{\sqrt{\alpha_3}}\right]\beta_n\right)\cos\left(\left[\frac{z_2-z_1}{\sqrt{\alpha_2}}\right]\beta_n\right)+\gamma_3\beta_n\cos\left(\left[\frac{z_3-z_2}{\sqrt{\alpha_3}}\right]\beta_n\right)\sin\left(\left[\frac{z_2-z_1}{\sqrt{\alpha_2}}\right]\beta_n\right) \\ &= \frac{\beta_n}{2}[\gamma_3+\gamma_2]\sin\left(\left[\frac{z_3-z_2}{\sqrt{\alpha_3}}+\frac{z_2-z_1}{\sqrt{\alpha_2}}\right]\beta_n\right)-\frac{\beta_n}{2}[\gamma_3-\gamma_2]\sin\left(\left[\frac{z_3-z_2}{\sqrt{\alpha_3}}-\frac{z_2-z_1}{\sqrt{\alpha_2}}\right]\beta_n\right) \end{aligned} \tag{6-84}$$

and the determinant of the matrix with appropriate columns replaced by the solution vector are given by the following:

Replacing column 1 with the solution vector:

$$\det\begin{vmatrix} \cos\left(\frac{z_1}{\sqrt{\alpha_1}}\beta_n\right) & -\cos\left(\frac{z_1}{\sqrt{\alpha_2}}\beta_n\right) & 0 & 0 \\ 0 & -\gamma_2\beta_n\sin\left(\frac{z_2}{\sqrt{\alpha_2}}\beta_n\right) & -\gamma_3\beta_n\cos\left(\frac{z_2}{\sqrt{\alpha_3}}\beta_n\right) & \gamma_3\beta_n\sin\left(\frac{z_2}{\sqrt{\alpha_3}}\beta_n\right) \\ 0 & \cos\left(\frac{z_2}{\sqrt{\alpha_2}}\beta_n\right) & -\sin\left(\frac{z_2}{\sqrt{\alpha_3}}\beta_n\right) & -\cos\left(\frac{z_2}{\sqrt{\alpha_3}}\beta_n\right) \\ 0 & 0 & \sin\left(\frac{z_3}{\sqrt{\alpha_3}}\beta_n\right) & \cos\left(\frac{z_3}{\sqrt{\alpha_3}}\beta_n\right) \end{vmatrix}$$

$$= -\frac{\beta_n}{2}\cos\left(\frac{z_1}{\sqrt{\alpha_1}}\beta_n\right)\left\{[\gamma_3+\gamma_2]\cos\left(\beta_n\left[\frac{z_2}{\sqrt{\alpha_2}}+\frac{z_3-z_2}{\sqrt{\alpha_3}}\right]\right)+[\gamma_3-\gamma_2]\cos\left(\beta_n\left[\frac{z_2}{\sqrt{\alpha_2}}-\frac{z_3-z_2}{\sqrt{\alpha_3}}\right]\right)\right\}$$

(6-85)

Replacing column 2 with the solution vector:

$$\det\begin{vmatrix} -\sin\left(\frac{z_1}{\sqrt{\alpha_2}}\beta_n\right) & -\cos\left(\frac{z_1}{\sqrt{\alpha_1}}\beta_n\right) & 0 & 0 \\ \gamma_2\beta_n\cos\left(\frac{z_2}{\sqrt{\alpha_2}}\beta_n\right) & 0 & -\gamma_3\beta_n\cos\left(\frac{z_2}{\sqrt{\alpha_3}}\beta_n\right) & \gamma_3\beta_n\sin\left(\frac{z_2}{\sqrt{\alpha_3}}\beta_n\right) \\ \sin\left(\frac{z_2}{\sqrt{\alpha_2}}\beta_n\right) & 0 & -\sin\left(\frac{z_2}{\sqrt{\alpha_3}}\beta_n\right) & -\cos\left(\frac{z_2}{\sqrt{\alpha_3}}\beta_n\right) \\ 0 & 0 & \sin\left(\frac{z_3}{\sqrt{\alpha_3}}\beta_n\right) & \cos\left(\frac{z_3}{\sqrt{\alpha_3}}\beta_n\right) \end{vmatrix}$$

$$= \frac{\beta_n}{2}\cos\left(\frac{z_1}{\sqrt{\alpha_1}}\beta_n\right)\left\{[\gamma_3+\gamma_2]\sin\left(\beta_n\left[\frac{z_2}{\sqrt{\alpha_2}}+\frac{z_3-z_2}{\sqrt{\alpha_3}}\right]\right)+[\gamma_3-\gamma_2]\cos\left(\beta_n\left[\frac{z_2}{\sqrt{\alpha_2}}-\frac{z_3-z_2}{\sqrt{\alpha_3}}\right]\right)\right\}$$

(6-86)

Replacing column 3 with the solution vector:

$$\det\begin{vmatrix} -\sin\left(\frac{z_1}{\sqrt{\alpha_2}}\beta_n\right) & -\cos\left(\frac{z_1}{\sqrt{\alpha_2}}\beta_n\right) & -\cos\left(\frac{z_1}{\sqrt{\alpha_1}}\beta_n\right) & 0 \\ \gamma_2\beta_n\cos\left(\frac{z_2}{\sqrt{\alpha_2}}\beta_n\right) & -\gamma_2\beta_n\sin\left(\frac{z_2}{\sqrt{\alpha_2}}\beta_n\right) & 0 & \gamma_3\beta_n\sin\left(\frac{z_2}{\sqrt{\alpha_3}}\beta_n\right) \\ \sin\left(\frac{z_2}{\sqrt{\alpha_2}}\beta_n\right) & \cos\left(\frac{z_2}{\sqrt{\alpha_2}}\beta_n\right) & 0 & -\cos\left(\frac{z_2}{\sqrt{\alpha_3}}\beta_n\right) \\ 0 & 0 & 0 & \cos\left(\frac{z_3}{\sqrt{\alpha_3}}\beta_n\right) \end{vmatrix}$$

$$= -\gamma_2\beta_n\cos\left(\frac{z_1}{\sqrt{\alpha_1}}\beta_n\right)\cos\left(\frac{z_3}{\sqrt{\alpha_3}}\beta_n\right)$$

(6-87)

and replacing column 4 with the solution vector:

$$\det\begin{vmatrix} -\sin\left(\frac{z_1}{\sqrt{\alpha_2}}\beta_n\right) & -\cos\left(\frac{z_1}{\sqrt{\alpha_2}}\beta_n\right) & 0 & -\cos\left(\frac{z_1}{\sqrt{\alpha_1}}\beta_n\right) \\ \gamma_2\beta_n\cos\left(\frac{z_2}{\sqrt{\alpha_2}}\beta_n\right) & -\gamma_2\beta_n\sin\left(\frac{z_2}{\sqrt{\alpha_2}}\beta_n\right) & -\gamma_3\beta_n\cos\left(\frac{z_2}{\sqrt{\alpha_3}}\beta_n\right) & 0 \\ \sin\left(\frac{z_2}{\sqrt{\alpha_2}}\beta_n\right) & \cos\left(\frac{z_2}{\sqrt{\alpha_2}}\beta_n\right) & -\sin\left(\frac{z_2}{\sqrt{\alpha_3}}\beta_n\right) & 0 \\ 0 & 0 & \sin\left(\frac{z_3}{\sqrt{\alpha_3}}\beta_n\right) & 0 \end{vmatrix}$$

$$= \gamma_2\beta_n\cos\left(\frac{z_1}{\sqrt{\alpha_1}}\beta_n\right)\sin\left(\frac{z_3}{\sqrt{\alpha_3}}\beta_n\right) \tag{6-88}$$

The positive roots of the system's characteristic equation result in the eigenvalues. The characteristic equation is given by setting the determinant of the matrix **A** equal to zero,

$$\det|\mathbf{A}| = 0 \tag{6-89}$$

where, after much simplification, the characteristic equation may be expressed as

$$\begin{aligned} &\left[\gamma_1\gamma_2+\gamma_1\gamma_3+\gamma_2^2+\gamma_2\gamma_3\right]\cos\left(\beta_n\left[\frac{z_3-z_2}{\sqrt{\alpha_3}}+\frac{z_2-z_1}{\sqrt{\alpha_2}}+\frac{z_1}{\sqrt{\alpha_1}}\right]\right)\ldots \\ &\ldots+\left[-\gamma_1\gamma_2-\gamma_1\gamma_3+\gamma_2^2+\gamma_2\gamma_3\right]\cos\left(\beta_n\left[\frac{z_3-z_2}{\sqrt{\alpha_3}}+\frac{z_2-z_1}{\sqrt{\alpha_2}}-\frac{z_1}{\sqrt{\alpha_1}}\right]\right)\ldots \\ &\ldots+\left[\gamma_1\gamma_2-\gamma_1\gamma_3-\gamma_2^2+\gamma_2\gamma_3\right]\cos\left(\beta_n\left[\frac{z_3-z_2}{\sqrt{\alpha_3}}-\frac{z_2-z_1}{\sqrt{\alpha_2}}+\frac{z_1}{\sqrt{\alpha_1}}\right]\right)\ldots \\ &\ldots+\left[-\gamma_1\gamma_2+\gamma_1\gamma_3-\gamma_2^2+\gamma_2\gamma_3\right]\cos\left(\beta_n\left[\frac{z_3-z_2}{\sqrt{\alpha_3}}-\frac{z_2-z_1}{\sqrt{\alpha_2}}-\frac{z_1}{\sqrt{\alpha_1}}\right]\right)=0 \end{aligned} \tag{6-90}$$

Hence the solution to the spatial problem takes the general form given in equation (6-78), where the coefficients are given by the matrix determinants calculated in equations (6-84) to (6-88). These solutions only exist for eigenvalues given by the positive roots of equation (6-90). When β is not an eigenvalue the problem has only a trivial solution, that is, $\Psi_i(\beta,x)=0$.

6.4.3.2 *Case 2 eigenvalue problem*

Matrix **A** for the transformed boundary conditions of equation (6-80) is given below, with $\gamma_i = \sqrt{D_{A(eff),i} \cdot R_{f,i}}$

$$\mathbf{A} = \begin{bmatrix} \frac{\beta_n}{\sqrt{\alpha_1}} & 0 & 0 & 0 & 0 & 0 \\ \gamma_1\beta_n \cos\left(\frac{z_1}{\sqrt{\alpha_1}}\beta_n\right) & -\gamma_1\beta_n \sin\left(\frac{z_1}{\sqrt{\alpha_1}}\beta_n\right) & -\gamma_2\beta_n \cos\left(\frac{z_1}{\sqrt{\alpha_2}}\beta_n\right) & \gamma_2\beta_n \sin\left(\frac{z_1}{\sqrt{\alpha_2}}\beta_n\right) & 0 & 0 \\ \sin\left(\frac{z_1}{\sqrt{\alpha_1}}\beta_n\right) & \cos\left(\frac{z_1}{\sqrt{\alpha_1}}\beta_n\right) & -\sin\left(\frac{z_1}{\sqrt{\alpha_2}}\beta_n\right) & -\cos\left(\frac{z_1}{\sqrt{\alpha_2}}\beta_n\right) & 0 & 0 \\ 0 & 0 & \gamma_2\beta_n \cos\left(\frac{z_2}{\sqrt{\alpha_2}}\beta_n\right) & -\gamma_2\beta_n \sin\left(\frac{z_2}{\sqrt{\alpha_2}}\beta_n\right) & -\gamma_3\beta_n \cos\left(\frac{z_2}{\sqrt{\alpha_3}}\beta_n\right) & \gamma_3\beta_n \sin\left(\frac{z_2}{\sqrt{\alpha_3}}\beta_n\right) \\ 0 & 0 & \sin\left(\frac{z_2}{\sqrt{\alpha_2}}\beta_n\right) & \cos\left(\frac{z_2}{\sqrt{\alpha_2}}\beta_n\right) & -\sin\left(\frac{z_2}{\sqrt{\alpha_3}}\beta_n\right) & -\cos\left(\frac{z_2}{\sqrt{\alpha_3}}\beta_n\right) \\ 0 & 0 & 0 & 0 & \left\{\begin{array}{l} k_a \sin\left(\frac{z_3}{\sqrt{\alpha_3}}\beta_n\right) \\ +\gamma_3\beta_n \cos\left(\frac{z_3}{\sqrt{\alpha_3}}\beta_n\right) \end{array}\right\} & \left\{\begin{array}{l} k_a \cos\left(\frac{z_3}{\sqrt{\alpha_3}}\beta_n\right) \\ -\gamma_3\beta_n \sin\left(\frac{z_3}{\sqrt{\alpha_3}}\beta_n\right) \end{array}\right\} \end{bmatrix} \tag{6-91}$$

Solving for the coefficients using the transformed boundary conditions, it may be seen from the coefficients of **A** that $A_{1,n} = 0$. Thus $B_{1,n}$ may be set to any arbitrary constant. Let $B_{1,n} = 1$, thus the remaining coefficients are given by

$$\mathbf{B}\cdot\begin{bmatrix} A_{2,n} \\ B_{2,n} \\ A_{3,n} \\ B_{3,n} \end{bmatrix} = \begin{bmatrix} -\cos\left(\frac{z_1}{\sqrt{\alpha_1}}\beta_n\right) \\ 0 \\ 0 \\ 0 \end{bmatrix} \tag{6-92}$$

where the matrix **B** is given by

$$\mathbf{B} = \begin{bmatrix} -\sin\left(\frac{z_1}{\sqrt{\alpha_2}}\beta_n\right) & -\cos\left(\frac{z_1}{\sqrt{\alpha_2}}\beta_n\right) & 0 & 0 \\ \gamma_2\beta_n\cos\left(\frac{z_2}{\sqrt{\alpha_2}}\beta_n\right) & -\gamma_2\beta_n\sin\left(\frac{z_2}{\sqrt{\alpha_2}}\beta_n\right) & -\gamma_3\beta_n\cos\left(\frac{z_2}{\sqrt{\alpha_3}}\beta_n\right) & \gamma_3\beta_n\sin\left(\frac{z_2}{\sqrt{\alpha_3}}\beta_n\right) \\ \sin\left(\frac{z_2}{\sqrt{\alpha_2}}\beta_n\right) & \cos\left(\frac{z_2}{\sqrt{\alpha_2}}\beta_n\right) & -\sin\left(\frac{z_2}{\sqrt{\alpha_3}}\beta_n\right) & -\cos\left(\frac{z_2}{\sqrt{\alpha_3}}\beta_n\right) \\ 0 & 0 & \left\{\begin{array}{l} k_a\sin\left(\frac{z_3}{\sqrt{\alpha_3}}\beta_n\right) \\ +\gamma_3\beta_n\cos\left(\frac{z_3}{\sqrt{\alpha_3}}\beta_n\right) \end{array}\right\} & \left\{\begin{array}{l} k_a\cos\left(\frac{z_3}{\sqrt{\alpha_3}}\beta_n\right) \\ -\gamma_3\beta_n\sin\left(\frac{z_3}{\sqrt{\alpha_3}}\beta_n\right) \end{array}\right\} \end{bmatrix} \tag{6-93}$$

Solving (6-92) for the coefficients using Cramer's rule, the determinant of the matrix **B** is

$$\begin{aligned} \Delta = \gamma_3\beta_n^{\,2}&\left[\gamma_2\cos\left(\frac{z_3-z_2}{\sqrt{\alpha_3}}\beta_n\right)\cos\left(\frac{z_2-z_1}{\sqrt{\alpha_2}}\beta_n\right)-\gamma_3\sin\left(\frac{z_3-z_2}{\sqrt{\alpha_3}}\beta_n\right)\sin\left(\frac{z_2-z_1}{\sqrt{\alpha_2}}\beta_n\right)\right] \\ +k_a\beta_n&\left[\gamma_2\sin\left(\frac{z_3-z_2}{\sqrt{\alpha_3}}\beta_n\right)\cos\left(\frac{z_2-z_1}{\sqrt{\alpha_2}}\beta_n\right)+\gamma_3\cos\left(\frac{z_3-z_2}{\sqrt{\alpha_3}}\beta_n\right)\sin\left(\frac{z_2-z_1}{\sqrt{\alpha_2}}\beta_n\right)\right] \end{aligned} \tag{6-94}$$

and the determinant of the matrix with appropriate columns replaced by the solution vector are given by the following:

Replacing column 1 with the solution vector:

$$
\det\begin{vmatrix}
-\cos\left(\frac{z_1}{\sqrt{\alpha_1}}\beta_n\right) & -\cos\left(\frac{z_1}{\sqrt{\alpha_2}}\beta_n\right) & 0 & 0 \\
0 & -\gamma_2\beta_n\sin\left(\frac{z_2}{\sqrt{\alpha_2}}\beta_n\right) & -\gamma_3\beta_n\cos\left(\frac{z_2}{\sqrt{\alpha_3}}\beta_n\right) & \gamma_3\beta_n\sin\left(\frac{z_2}{\sqrt{\alpha_3}}\beta_n\right) \\
0 & \cos\left(\frac{z_2}{\sqrt{\alpha_2}}\beta_n\right) & -\sin\left(\frac{z_2}{\sqrt{\alpha_3}}\beta_n\right) & -\cos\left(\frac{z_2}{\sqrt{\alpha_3}}\beta_n\right) \\
0 & 0 & \left\{\begin{matrix} k_a\sin\left(\frac{z_3}{\sqrt{\alpha_3}}\beta_n\right) \\ +\gamma_3\beta_n\cos\left(\frac{z_3}{\sqrt{\alpha_3}}\beta_n\right)\end{matrix}\right\} & \left\{\begin{matrix} k_a\cos\left(\frac{z_3}{\sqrt{\alpha_3}}\beta_n\right) \\ -\gamma_3\beta_n\sin\left(\frac{z_3}{\sqrt{\alpha_3}}\beta_n\right)\end{matrix}\right\}
\end{vmatrix}
$$

$$
= k_a\beta_n\cos\left(\frac{z_1}{\sqrt{\alpha_1}}\beta_n\right)\left\{\gamma_2\sin\left(\frac{z_2}{\sqrt{\alpha_2}}\beta_n\right)\sin\left(\frac{z_3-z_2}{\sqrt{\alpha_3}}\beta_n\right)-\gamma_3\cos\left(\frac{z_2}{\sqrt{\alpha_2}}\beta_n\right)\cos\left(\frac{z_3-z_2}{\sqrt{\alpha_3}}\beta_n\right)\right\}
$$

$$
+\gamma_3\beta_n^2\cos\left(\frac{z_1}{\sqrt{\alpha_1}}\beta_n\right)\left\{\gamma_2\sin\left(\frac{z_2}{\sqrt{\alpha_2}}\beta_n\right)\cos\left(\frac{z_3-z_2}{\sqrt{\alpha_3}}\beta_n\right)+\gamma_3\cos\left(\frac{z_2}{\sqrt{\alpha_2}}\beta_n\right)\sin\left(\frac{z_3-z_2}{\sqrt{\alpha_3}}\beta_n\right)\right\}
$$

(6-95)

Replacing column 2 with the solution vector:

$$
\det\begin{vmatrix}
-\sin\left(\frac{z_1}{\sqrt{\alpha_2}}\beta_n\right) & -\cos\left(\frac{z_1}{\sqrt{\alpha_1}}\beta_n\right) & 0 & 0 \\
\gamma_2\beta_n\cos\left(\frac{z_2}{\sqrt{\alpha_2}}\beta_n\right) & 0 & -\gamma_3\beta_n\cos\left(\frac{z_2}{\sqrt{\alpha_3}}\beta_n\right) & \gamma_3\beta_n\sin\left(\frac{z_2}{\sqrt{\alpha_3}}\beta_n\right) \\
\sin\left(\frac{z_2}{\sqrt{\alpha_2}}\beta_n\right) & 0 & -\sin\left(\frac{z_2}{\sqrt{\alpha_3}}\beta_n\right) & -\cos\left(\frac{z_2}{\sqrt{\alpha_3}}\beta_n\right) \\
0 & 0 & \left\{\begin{matrix} k_a\sin\left(\frac{z_3}{\sqrt{\alpha_3}}\beta_n\right) \\ +\gamma_3\beta_n\cos\left(\frac{z_3}{\sqrt{\alpha_3}}\beta_n\right)\end{matrix}\right\} & \left\{\begin{matrix} k_a\cos\left(\frac{z_3}{\sqrt{\alpha_3}}\beta_n\right) \\ -\gamma_3\beta_n\sin\left(\frac{z_3}{\sqrt{\alpha_3}}\beta_n\right)\end{matrix}\right\}
\end{vmatrix}
$$

$$
= k_a\beta_n\cos\left(\frac{z_1}{\sqrt{\alpha_1}}\beta_n\right)\left\{\gamma_2\cos\left(\frac{z_2}{\sqrt{\alpha_2}}\beta_n\right)\sin\left(\frac{z_3-z_2}{\sqrt{\alpha_3}}\beta_n\right)+\gamma_3\sin\left(\frac{z_2}{\sqrt{\alpha_2}}\beta_n\right)\cos\left(\frac{z_3-z_2}{\sqrt{\alpha_3}}\beta_n\right)\right\}
$$

$$
+\gamma_3\beta_n^2\cos\left(\frac{z_1}{\sqrt{\alpha_1}}\beta_n\right)\left\{\gamma_2\cos\left(\frac{z_2}{\sqrt{\alpha_2}}\beta_n\right)\cos\left(\frac{z_3-z_2}{\sqrt{\alpha_3}}\beta_n\right)-\gamma_3\sin\left(\frac{z_2}{\sqrt{\alpha_2}}\beta_n\right)\sin\left(\frac{z_3-z_2}{\sqrt{\alpha_3}}\beta_n\right)\right\}
$$

(6-96)

Replacing column 3 with the solution vector:

$$\det\begin{vmatrix} -\sin\left(\frac{z_1}{\sqrt{\alpha_2}}\beta_n\right) & -\cos\left(\frac{z_1}{\sqrt{\alpha_2}}\beta_n\right) & -\cos\left(\frac{z_1}{\sqrt{\alpha_1}}\beta_n\right) & 0 \\ \gamma_2\beta_n\cos\left(\frac{z_2}{\sqrt{\alpha_2}}\beta_n\right) & -\gamma_2\beta_n\sin\left(\frac{z_2}{\sqrt{\alpha_2}}\beta_n\right) & 0 & \gamma_3\beta_n\sin\left(\frac{z_2}{\sqrt{\alpha_3}}\beta_n\right) \\ \sin\left(\frac{z_2}{\sqrt{\alpha_2}}\beta_n\right) & \cos\left(\frac{z_2}{\sqrt{\alpha_2}}\beta_n\right) & 0 & -\cos\left(\frac{z_2}{\sqrt{\alpha_3}}\beta_n\right) \\ 0 & 0 & 0 & \left\{\begin{matrix} k_a\cos\left(\frac{z_3}{\sqrt{\alpha_3}}\beta_n\right) \\ -\gamma_3\beta_n\sin\left(\frac{z_3}{\sqrt{\alpha_3}}\beta_n\right)\end{matrix}\right\} \end{vmatrix}$$

$$= -k_a\gamma_2\beta_n\cos\left(\frac{z_1}{\sqrt{\alpha_1}}\beta_n\right)\cos\left(\frac{z_3}{\sqrt{\alpha_3}}\beta_n\right) + \gamma_2\gamma_3\beta_n^2\cos\left(\frac{z_1}{\sqrt{\alpha_1}}\beta_n\right)\sin\left(\frac{z_3}{\sqrt{\alpha_3}}\beta_n\right)$$

(6-97)

Replacing column 4 with the solution vector:

$$\det\begin{vmatrix} -\sin\left(\frac{z_1}{\sqrt{\alpha_2}}\beta_n\right) & -\cos\left(\frac{z_1}{\sqrt{\alpha_2}}\beta_n\right) & 0 & -\cos\left(\frac{z_1}{\sqrt{\alpha_1}}\beta_n\right) \\ \gamma_2\beta_n\cos\left(\frac{z_2}{\sqrt{\alpha_2}}\beta_n\right) & -\gamma_2\beta_n\sin\left(\frac{z_2}{\sqrt{\alpha_2}}\beta_n\right) & -\gamma_3\beta_n\cos\left(\frac{z_2}{\sqrt{\alpha_3}}\beta_n\right) & 0 \\ \sin\left(\frac{z_2}{\sqrt{\alpha_2}}\beta_n\right) & \cos\left(\frac{z_2}{\sqrt{\alpha_2}}\beta_n\right) & -\sin\left(\frac{z_2}{\sqrt{\alpha_3}}\beta_n\right) & 0 \\ 0 & 0 & \left\{\begin{matrix} k_a\sin\left(\frac{z_3}{\sqrt{\alpha_3}}\beta_n\right) \\ +\gamma_3\beta_n\cos\left(\frac{z_3}{\sqrt{\alpha_3}}\beta_n\right)\end{matrix}\right\} & 0 \end{vmatrix}$$

$$= k_a\gamma_2\beta_n\cos\left(\frac{z_1}{\sqrt{\alpha_1}}\beta_n\right)\sin\left(\frac{z_3}{\sqrt{\alpha_3}}\beta_n\right) + \gamma_2\gamma_3\beta_n^2\cos\left(\frac{z_1}{\sqrt{\alpha_1}}\beta_n\right)\cos\left(\frac{z_3}{\sqrt{\alpha_3}}\beta_n\right)$$

(6-98)

The positive roots of the system's characteristic equation result in the eigenvalues. The characteristic equation is given by setting the determinant of the matrix **A** equal to zero.

$$\det|\mathbf{A}| = 0 \tag{6-99}$$

where, after much simplification, the characteristic equation may be expressed as

$$\det|\mathbf{A}| = \tfrac{1}{4}k_a\beta_n^{\ 2}\left\{\begin{aligned}&\left[\gamma_1\gamma_2+\gamma_1\gamma_3+\gamma_2^{\ 2}+\gamma_2\gamma_3\right]\cos\left(\beta_n\left[\frac{z_3-z_2}{\sqrt{\alpha_3}}+\frac{z_2-z_1}{\sqrt{\alpha_2}}+\frac{z_1}{\sqrt{\alpha_1}}\right]\right)\\&+\left[-\gamma_1\gamma_2-\gamma_1\gamma_3+\gamma_2^{\ 2}+\gamma_2\gamma_3\right]\cos\left(\beta_n\left[\frac{z_3-z_2}{\sqrt{\alpha_3}}+\frac{z_2-z_1}{\sqrt{\alpha_2}}-\frac{z_1}{\sqrt{\alpha_1}}\right]\right)\\&+\left[\gamma_1\gamma_2-\gamma_1\gamma_3-\gamma_2^{\ 2}+\gamma_2\gamma_3\right]\cos\left(\beta_n\left[\frac{z_3-z_2}{\sqrt{\alpha_3}}-\frac{z_2-z_1}{\sqrt{\alpha_2}}+\frac{z_1}{\sqrt{\alpha_1}}\right]\right)\\&+\left[-\gamma_1\gamma_2+\gamma_1\gamma_3-\gamma_2^{\ 2}+\gamma_2\gamma_3\right]\cos\left(\beta_n\left[\frac{z_3-z_2}{\sqrt{\alpha_3}}-\frac{z_2-z_1}{\sqrt{\alpha_2}}-\frac{z_1}{\sqrt{\alpha_1}}\right]\right)\end{aligned}\right\}\cdots$$

$$\cdots+\tfrac{1}{4}\gamma_3\beta_n^{\ 3}\left\{\begin{aligned}&\left[-\gamma_1\gamma_2-\gamma_1\gamma_3-\gamma_2^{\ 2}-\gamma_2\gamma_3\right]\sin\left(\beta_n\left[\frac{z_3-z_2}{\sqrt{\alpha_3}}+\frac{z_2-z_1}{\sqrt{\alpha_2}}+\frac{z_1}{\sqrt{\alpha_1}}\right]\right)\\&+\left[\gamma_1\gamma_2+\gamma_1\gamma_3-\gamma_2^{\ 2}-\gamma_2\gamma_3\right]\sin\left(\beta_n\left[\frac{z_3-z_2}{\sqrt{\alpha_3}}+\frac{z_2-z_1}{\sqrt{\alpha_2}}-\frac{z_1}{\sqrt{\alpha_1}}\right]\right)\\&+\left[-\gamma_1\gamma_2+\gamma_1\gamma_3+\gamma_2^{\ 2}-\gamma_2\gamma_3\right]\sin\left(\beta_n\left[\frac{z_3-z_2}{\sqrt{\alpha_3}}-\frac{z_2-z_1}{\sqrt{\alpha_2}}+\frac{z_1}{\sqrt{\alpha_1}}\right]\right)\\&+\left[\gamma_1\gamma_2-\gamma_1\gamma_3+\gamma_2^{\ 2}-\gamma_2\gamma_3\right]\sin\left(\beta_n\left[\frac{z_3-z_2}{\sqrt{\alpha_3}}-\frac{z_2-z_1}{\sqrt{\alpha_2}}-\frac{z_1}{\sqrt{\alpha_1}}\right]\right)\end{aligned}\right\} \tag{6-100}$$

Hence the solution to the spatial problem takes the general form given in equation (6-78), where the coefficients are given by the matrix determinants calculated in equations (6-94) to (6-98). These solutions only exist for eigenvalues given by the positive roots of equation (6-100). When β is not an eigenvalue the problem has only a trivial solution, that is, $\Psi_i(\beta,x)=0$.

6.4.4 Initial conditions

The complete solution to the system, $c_{A,i}(z,t)$, in each layer $i=1,2,3$ is constructed by a linear superposition of product of the temporal and spatial solutions

$$c_{A,i}(z,t) = \sum_{n=1}^{\infty}\delta_n\cdot\Psi_i(\beta_n,z)\exp\left[-\beta_n^{\ 2}t\right] \tag{6-101}$$

where δ_n are constant coefficients used to satisfy the initial boundary conditions. The values of δ_n can be explicitly determined by solving equation (6-101) at $t=0$

$$c_{A,i}(z,t)\Big|_{t=0} = \sum_{n=1}^{\infty} \delta_n \cdot \Psi_i(\beta_n, z) = c_{A0}(z) \tag{6-102}$$

and using the property that the eigenfunctions are orthogonal

$$\begin{bmatrix} R_{f,1} \int_0^{z_1} \Psi_1(\beta_m, z) \cdot \Psi_1(\beta_n, z) dz \ldots \\ \ldots + R_{f,2} \int_{z_1}^{z_2} \Psi_2(\beta_m, z) \cdot \Psi_2(\beta_n, z) dz \ldots \\ \ldots + R_{f,3} \int_{z_2}^{z_3} \Psi_3(\beta_m, z) \cdot \Psi_3(\beta_n, z) dz \end{bmatrix} = \begin{cases} 0 & \text{for } m \neq n \\ N(\beta_n) & \text{for } m = n \end{cases} \tag{6-103}$$

where the *normalization integral*, $N(\beta_n)$, is defined as by the following expression:

$$N(\beta_n) = \sum_{i=1}^{3} \left\{ R_{f,i} \int_{z_{i-1}}^{z_i} [\Psi_i(\beta_n, z)]^2 dz \right\} \tag{6-104}$$

where $z_0 = 0$.

Operate both sides of equation (6-102) with the operator $R_{f,i} \int_{z_{i-1}}^{z_i} \Psi_i(\beta_r, z) dz$ for an arbitrary r, and summing up for all $i = 1,2,3$ we obtain

$$\sum_{n=1}^{\infty} \delta_n \left[\sum_{i=1}^{3} R_{f,i} \int_{z_{i-1}}^{z_i} \Psi_i(\beta_r, z) \cdot \Psi_i(\beta_n, z) dz \right] = \sum_{i=1}^{3} R_{f,i} \int_{z_{i-1}}^{z_i} \Psi_i(\beta_r, z) \cdot c_{A0}(z) dz \tag{6-105}$$

The right-hand side of the above equation is the *initialization integral*, $I_0(\beta_n)$, as defined by equation (6-15). The left-hand side can be simplified using the orthogonality property given in equation (6-103). Thus we have nonzero values only when $r = n$, hence

$$\delta_n \cdot N(\beta_n) = I_0(\beta_n) \tag{6-106}$$

Substituting for the constant coefficient, the generalized solution for the concentration thus is given by

$$c_{A,i}(z,t) = \sum_{n=1}^{\infty} \frac{1}{N(\beta_n)} \cdot I_0(\beta_n) \cdot \Psi_i(\beta_n, z) \exp\left[-\beta_n^2 t\right] \tag{6-107}$$

Differentiating equation (6-107) at $z = z_3$ gives a general solution for the surface flux

$$j_A(t)\big|_{z=L} = -D_{A(eff),3}\frac{\partial c_A}{\partial z}\bigg|_{z=z_3} = -D_{A(eff),3}\sum_{n=1}^{\infty}\left\{\frac{1}{N(\beta_n)}\cdot I_0(\beta_n)\cdot\frac{\partial\Psi_3}{\partial z}\bigg|_{z=z_3}\exp\left[-\beta_n^2 t\right]\right\} \tag{6-108}$$

6.4.5 Variable transformation for first-order decay

Solving the system with the inclusion of a first-order decay of the contaminant species, as described by the dynamic equations (6-16), (6-17), and (6-18), we use the following variable substitution:

$$c_{A,i}(z,t) = \hat{c}_i(z,t)\exp(-k_1 t) \tag{6-109}$$

Effect of transformation on the dynamic equation

The transformed left-hand side of equations (6-16), (6-17), and (6-18), the partial derivative with respect to time, gives

$$\frac{\partial c_{A,i}}{\partial t} = -k_1\hat{c}_i(z,t)\exp(-k_1 t) + \frac{\partial\hat{c}_i}{\partial t}\exp(-k_1 t) \tag{6-110}$$

The transformed right-hand side, spatial terms, gives

$$\left(\frac{D_{A(eff),i}}{R_{f.i}}\right)\frac{\partial^2 c_{A,i}}{\partial z^2} - k_1 c_{A,i}(z,t) = \left(\frac{D_{A(eff),i}}{R_{f,i}}\right)\frac{\partial^2\hat{c}_i}{\partial z^2}\exp(-k_1 t) - k_1\hat{c}_i(z,t)\exp(-k_1 t) \tag{6-111}$$

Hence, the transformed dynamic equations result in

$$\therefore\quad \frac{\partial\hat{c}_i}{\partial t} = \left(\frac{D_{A(eff),i}}{R_{f,i}}\right)\frac{\partial^2\hat{c}_i}{\partial z^2} \tag{6-112}$$

The net effect of the transformation is to remove the reaction term from the dynamic equation, resulting in the standard diffusion-adsorption formulation.

Effect of transformation on initial conditions

At time equal to zero the transformed initial concentration results in

$$c_{A,i}(z,0) = \hat{c}_i(z,0)\exp(0) \equiv \hat{c}_i(z,0) \tag{6-113}$$

Hence, all initial conditions remain unaltered after the transformation.

Effect of the transformation on boundary conditions

Transformation of the boundary conditions at the interface and the bottom boundary results in

$$\hat{c}_1(z,t)\exp(-k_1 t) = \hat{c}_2(z,t)\exp(-k_1 t) \qquad \text{at } z = z_1,\ t > 0 \tag{6-114}$$

$$D_{A(eff),1}\frac{\partial\hat{c}_1}{\partial z}\exp(-k_1 t) = D_{A(eff),2}\frac{\partial\hat{c}_2}{\partial z}\exp(-k_1 t) \qquad \text{at } z = z_1,\ t > 0 \tag{6-115}$$

$$\hat{c}_2(z,t)\exp(-k_1 t) = \hat{c}_3(z,t)\exp(-k_1 t) \qquad \text{at } z = z_2,\ t > 0 \tag{6-116}$$

$$D_{A(eff),2}\frac{\partial \hat{c}_2}{\partial z}\exp(-k_1 t) = D_{A(eff),3}\frac{\partial \hat{c}_3}{\partial z}\exp(-k_1 t) \qquad \text{at } z = z_2,\ t > 0 \quad (6\text{-}117)$$

$$\frac{\partial \hat{c}_1}{\partial z}\exp(-k_1 t) = 0 \qquad \text{at } z = 0,\ t > 0 \quad (6\text{-}118)$$

Transformation of the upper boundary condition for the case of zero surface concentration gives

$$\hat{c}_3(z,t)\exp(-k_1 t) = 0 \qquad \text{at } z = z_3,\ t > 0 \quad (6\text{-}119)$$

and for a mass transfer upper boundary gives

$$D_{A(eff),3}\frac{\partial \hat{c}_3}{\partial z}\exp(-k_1 t) + k_a \cdot \hat{c}_3(z,t)\exp(-k_1 t) = 0 \qquad \text{at } z = z_3,\ t > 0 \quad (6\text{-}120)$$

Dividing all these equations by the common factor $\exp(-k_1 t)$ shows that the boundary conditions remain unaltered after the transformation.

Effect on the solution and surface flux

As the transformed system of equations results in the standard formulation with diffusion and adsorption mechanisms only, we solve for $\hat{c}_i(z,t)$ as described in the pervious sections of 6.4. The resulting solution then is inverted back into terms of $c_{A,i}(z,t)$ by equation (6-109).

The net effect of the decay term is to include an additional exponential time term to the solution. Since the flux is determined by a differential with respect to space, it too is only modified by this additional factor.

References

Özisk, M.N. (1993) *Heat Conduction*, 2nd ed., John Wiley & Sons, New York.

Carslaw, H.S., Jaeger, J.C. (1959) *Conduction of Heat in Solids*, Clarendon Press, Oxford.

Thoma, G., Popov, V., Bowen, M. (1997) Multi-layer Caps for Containment of Contaminated Sediments, *WERC/HSRC Joint Conference on the Environment Proceedings*, WERC New Mexico State University, NM.

7 Advection-diffusion models

7.1 Introduction

In many environmental situations it is not possible to neglect advection, even for vertical transport through multilayered media. The contaminated mobile phase may be subject to advective forces (i.e., moving groundwater) in addition to basic diffusive/dispersive forces. The advection-diffusion equation may also be used to model experiments (i.e., columns) which are used in determining important properties of the sediment being studied.

van Genuchten and Alves (1982) have provided a compendium of many analytical solutions to the advection-diffusion equation. Several of these solutions are summarized in Section 7.2. A brief discussion on the numerical evaluation of the solutions is given in Section 7.3. The system equations were solved using a Laplace transform technique that is outlined in 7.4. Although the primary focus in this volume is on diffusive problems, selected advection-diffusion models are included for completeness.

7.2 Analysis summary

The general form of the advection-diffusion equation is given by the equation

$$R_f \frac{\partial c_A}{\partial t} = D_A \frac{\partial^2 c_A}{\partial z^2} - v \frac{\partial c_A}{\partial z} \qquad z \in [0,\infty) \tag{7-1}$$

where the diffusive component is described by the second-order partial derivative of concentration multiplied by the diffusion/dispersion coefficient, and the advective component by the first-order partial derivative multiplied by the average fluid velocity. The relative magnitude of the advective process compared to the diffusive process is often represented by the dimensionless variable, the Peclet number

$$\text{Pe} = \frac{v \cdot L_{char}}{D_A} \tag{7-2}$$

where L_{char} is the characteristic length of the system (e.g., column length) or contaminated layer depth.

7.2.1 Case 1: Semi-infinite region with uniform initial concentration with a constant concentration boundary condition

Initial and boundary conditions

$$c_A(z,t)\big|_{t=0} = c_i \qquad z \in [0,\infty) \tag{7-3}$$

$$c_A(z,t)\big|_{z=0} = c_o \qquad t > 0 \tag{7-4}$$

$$\left.\frac{\partial c_A}{\partial z}\right|_{z\to\infty} = 0 \qquad t > 0 \tag{7-5}$$

Solution

$$c_A(z,t) = c_i + (c_o - c_i)\left[\frac{1}{2}\text{erfc}\left(\frac{R_f z - vt}{2\sqrt{D_A R_f t}}\right) + \frac{1}{2}\exp\left(\frac{vz}{D_A}\right)\text{erfc}\left(\frac{R_f z + vt}{2\sqrt{D_A R_f t}}\right)\right] \quad (7\text{-}6)$$

7.2.2 Case 2: Semi-infinite region with uniform initial concentration with a constant flux boundary condition

Initial and boundary conditions

$$c_A(z,t)\big|_{t=0} = c_i \qquad z \in [0,\infty) \quad (7\text{-}7)$$

$$-D_A \frac{\partial c_A}{\partial z}\bigg|_{z=0} + v \cdot c_A(z,t)\big|_{z=0} = v \cdot c_o \qquad t > 0 \quad (7\text{-}8)$$

$$\frac{\partial c_A}{\partial z}\bigg|_{z\to\infty} = 0 \qquad t > 0 \quad (7\text{-}9)$$

Solution

$$c_A(z,t) = c_i + (c_o - c_i)\left[\begin{array}{l} \frac{1}{2}\text{erfc}\left(\frac{R_f z - vt}{2\sqrt{D_A R_f t}}\right) + \sqrt{\frac{v^2 t}{\pi D_A R_f}}\exp\left(-\frac{(R_f z - vt)^2}{4 D_A R_f t}\right)\ldots \\ \ldots - \frac{1}{2}\left(1 + \frac{vz}{D_A} + \frac{v^2 t}{D_A R_f}\right)\exp\left(\frac{vz}{D_A}\right)\text{erfc}\left(\frac{R_f z + vt}{2\sqrt{D_A R_f t}}\right) \end{array}\right] \quad (7\text{-}10)$$

7.2.3 Case 3: Semi-infinite region with uniform initial concentration with a boundary condition given by a finite-timed pulse at a constant concentration

Initial and boundary conditions

$$c_A(z,t)\big|_{t=0} = c_i \qquad z \in [0,\infty) \quad (7\text{-}11)$$

$$c_A(z,t)\big|_{z=0} = \begin{cases} c_o & t > 0 \\ 0 & t > t_0 \end{cases} \quad (7\text{-}12)$$

$$\frac{\partial c_A}{\partial z}\bigg|_{z\to\infty} = 0 \qquad t > 0 \quad (7\text{-}13)$$

Solution

$$c_A(z,t) = c_i + (c_o - c_i)\left[\frac{1}{2}\operatorname{erfc}\left(\frac{R_f z - vt}{2\sqrt{D_A R_f t}}\right) + \frac{1}{2}\exp\left(\frac{vz}{D_A}\right)\operatorname{erfc}\left(\frac{R_f z + vt}{2\sqrt{D_A R_f t}}\right)\right] \dots$$

$$\dots - c_o\left[\frac{1}{2}\operatorname{erfc}\left(\frac{R_f z - v(t-t_0)}{2\sqrt{D_A R_f (t-t_0)}}\right) + \frac{1}{2}\exp\left(\frac{vz}{D_A}\right)\operatorname{erfc}\left(\frac{R_f z + v(t-t_0)}{2\sqrt{D_A R_f (t-t_0)}}\right)\right] \tag{7-14}$$

7.2.4 Case 4: Semi-infinite region with uniform initial concentration with a boundary condition given by a finite-timed pulse at a constant flux

Initial and boundary conditions

$$c_A(z,t)\big|_{t=0} = c_i \qquad z \in [0,\infty) \tag{7-15}$$

$$-D_A \frac{\partial c_A}{\partial z}\bigg|_{z=0} + v \cdot c_A(z,t)\big|_{z=0} = \begin{cases} v \cdot c_o & t > 0 \\ 0 & t > t_0 \end{cases} \tag{7-16}$$

$$\frac{\partial c_A}{\partial z}\bigg|_{z\to\infty} = 0 \qquad t > 0 \tag{7-17}$$

Solution

$$c_A(z,t) = c_i + (c_o - c_i)\left[\begin{array}{l} \frac{1}{2}\operatorname{erfc}\left(\frac{R_f z - vt}{2\sqrt{D_A R_f t}}\right) + \sqrt{\frac{v^2 t}{\pi D_A R_f}}\exp\left(-\frac{(R_f z - vt)^2}{4 D_A R_f t}\right)\dots \\ \dots - \frac{1}{2}\left(1 + \frac{vz}{D_A} + \frac{v^2 t}{D_A R_f}\right)\exp\left(\frac{vz}{D_A}\right)\operatorname{erfc}\left(\frac{R_f z + vt}{2\sqrt{D_A R_f t}}\right) \end{array}\right]\dots$$

$$\dots - c_o\left[\begin{array}{l} \frac{1}{2}\operatorname{erfc}\left(\frac{R_f z - v(t-t_0)}{2\sqrt{D_A R_f (t-t_0)}}\right) + \sqrt{\frac{v^2 (t-t_0)}{\pi D_A R_f}}\exp\left(-\frac{(R_f z - v(t-t_0))^2}{4 D_A R_f (t-t_0)}\right)\dots \\ \dots - \frac{1}{2}\left(1 + \frac{vz}{D_A} + \frac{v^2 (t-t_0)}{D_A R_f}\right)\exp\left(\frac{vz}{D_A}\right)\operatorname{erfc}\left(\frac{R_f z + v(t-t_0)}{2\sqrt{D_A R_f (t-t_0)}}\right) \end{array}\right] \tag{7-18}$$

7.2.5 Case 5: Semi-infinite region with uniform initial concentration capped by a finite region of a different uniform initial condition, with a constant concentration boundary condition

Initial and boundary conditions

$$c_A(z,t)\big|_{t=0} = \begin{cases} c_1 & z \in [0, z_1) \\ c_2 & z \in [z_1, \infty) \end{cases} \tag{7-19}$$

$$c_A(z,t)\big|_{z=0} = c_o \qquad t>0 \tag{7-20}$$

$$\left.\frac{\partial c_A}{\partial z}\right|_{z\to\infty} = 0 \qquad t>0 \tag{7-21}$$

Solution

$$c_A(z,t) = c_2 + (c_1 - c_2)\cdot\psi(z,t) + (c_o - c_1)\cdot\phi(z,t) \tag{7-22}$$

where

$$\psi(z,t) = \frac{1}{2}\text{erfc}\left[\frac{R_f(z-z_1)-vt}{2\sqrt{D_A R_f t}}\right] + \frac{1}{2}\exp\left(\frac{vz}{D_A}\right)\text{erfc}\left[\frac{R_f(z+z_1)+vt}{2\sqrt{D_A R_f t}}\right] \tag{7-23}$$

$$\phi(z,t) = \frac{1}{2}\text{erfc}\left[\frac{R_f z - vt}{2\sqrt{D_A R_f t}}\right] + \frac{1}{2}\exp\left(\frac{vz}{D_A}\right)\text{erfc}\left[\frac{R_f z + vt}{2\sqrt{D_A R_f t}}\right] \tag{7-24}$$

7.2.6 Case 6: Semi-infinite region with uniform initial concentration capped by a finite region of a different uniform initial condition, with a constant flux boundary condition

Initial and boundary conditions

$$c_A(z,t)\big|_{t=0} = \begin{cases} c_1 & z\in[0,z_1) \\ c_2 & z\in[z_1,\infty) \end{cases} \tag{7-25}$$

$$-D_A\left.\frac{\partial c_A}{\partial z}\right|_{z=0} + v\cdot c_A(z,t)\big|_{z=0} = v\cdot c_o \qquad t>0 \tag{7-26}$$

$$\left.\frac{\partial c_A}{\partial z}\right|_{z\to\infty} = 0 \qquad t>0 \tag{7-27}$$

Solution

$$c_A(z,t) = c_2 + (c_1 - c_2)\cdot\eta(z,t) + (c_o - c_1)\cdot\chi(z,t) \tag{7-28}$$

where

$$\eta(z,t) = \frac{1}{2}\text{erfc}\left[\frac{R_f(z-z_1)-vt}{2\sqrt{D_A R_f t}}\right] + \sqrt{\frac{v^2 t}{\pi D_A R_f}}\exp\left[\frac{vx}{D_A} - \frac{R_f}{4D_A t}\left(z+z_1+\frac{vt}{R_f}\right)^2\right]\ldots$$
$$\ldots - \frac{1}{2}\left(1+\frac{v(z+z_1)}{D_A}+\frac{v^2 t}{D_A R_f}\right)\exp\left(\frac{vz}{D_A}\right)\text{erfc}\left[\frac{R_f(z+z_1)+vt}{2\sqrt{D_A R_f t}}\right] \tag{7-29}$$

$$\chi(z,t)=\frac{1}{2}\operatorname{erfc}\left[\frac{R_f z-vt}{2\sqrt{D_A R_f t}}\right]+\sqrt{\frac{v^2 t}{\pi D_A R_f}}\exp\left[-\frac{\left(R_f z-vt\right)^2}{4D_A R_f t}\right]\ldots$$
$$\ldots-\frac{1}{2}\left(1+\frac{vz}{D_A}+\frac{v^2 t}{D_A R_f}\right)\exp\left(\frac{vz}{D_A}\right)\operatorname{erfc}\left[\frac{R_f z+vt}{2\sqrt{D_A R_f t}}\right] \tag{7-30}$$

7.2.7 Case 7: Semi-infinite region with uniform initial concentration capped by a finite region of a different uniform initial condition, with a boundary condition given by a finite-timed pulse at a constant concentration

Initial and boundary conditions

$$c_A(z,t)\big|_{t=0}=\begin{cases}c_1 & z\in[0,z_1)\\ c_2 & z\in[z_1,\infty)\end{cases} \tag{7-31}$$

$$c_A(z,t)\big|_{z=0}=\begin{cases}c_o & t>0\\ 0 & t>t_0\end{cases} \tag{7-32}$$

$$\left.\frac{\partial c_A}{\partial z}\right|_{z\to\infty}=0 \qquad t>0 \tag{7-33}$$

Solution

$$c_A(z,t)=c_2+(c_1-c_2)\cdot\psi(z,t)+(c_o-c_1)\cdot\phi(z,t)-c_o\cdot\phi(z,t-t_0) \tag{7-34}$$

where

$$\psi(z,t)=\frac{1}{2}\operatorname{erfc}\left[\frac{R_f(z-z_1)-vt}{2\sqrt{D_A R_f t}}\right]+\frac{1}{2}\exp\left(\frac{vz}{D_A}\right)\operatorname{erfc}\left[\frac{R_f(z+z_1)+vt}{2\sqrt{D_A R_f t}}\right] \tag{7 35}$$

$$\phi(z,t)=\frac{1}{2}\operatorname{erfc}\left[\frac{R_f z-vt}{2\sqrt{D_A R_f t}}\right]+\frac{1}{2}\exp\left(\frac{vz}{D_A}\right)\operatorname{erfc}\left[\frac{R_f z+vt}{2\sqrt{D_A R_f t}}\right] \tag{7-36}$$

7.2.8 Case 8: Semi-infinite region with uniform initial concentration capped by a finite region of a different uniform initial condition, with a boundary condition given by a finite-timed pulse at a constant flux

Initial and boundary conditions

$$c_A(z,t)\big|_{t=0}=\begin{cases}c_1 & z\in[0,z_1)\\ c_2 & z\in[z_1,\infty)\end{cases} \tag{7-37}$$

$$-D_A\left.\frac{\partial c_A}{\partial z}\right|_{z=0}+v\cdot c_A(z,t)\big|_{z=0}=\begin{cases}v\cdot c_o & t>0\\ 0 & t>t_0\end{cases} \tag{7-38}$$

$$\left.\frac{\partial c_A}{\partial z}\right|_{z\to\infty} = 0 \qquad t > 0 \tag{7-39}$$

Solution

$$c_A(z,t) = c_2 + (c_1 - c_2)\cdot\eta(z,t) + (c_o - c_1)\cdot\chi(z,t) - c_o\cdot\chi(z,t) \tag{7-40}$$

where

$$\eta(z,t) = \frac{1}{2}\text{erfc}\left[\frac{R_f(z-z_1)-vt}{2\sqrt{D_A R_f t}}\right] + \sqrt{\frac{v^2 t}{\pi D_A R_f}}\exp\left[\frac{vx}{D_A} - \frac{R_f}{4D_A t}\left(z + z_1 + \frac{vt}{R_f}\right)^2\right]\ldots$$
$$\ldots - \frac{1}{2}\left(1 + \frac{v(z+z_1)}{D_A} + \frac{v^2 t}{D_A R_f}\right)\exp\left(\frac{vz}{D_A}\right)\text{erfc}\left[\frac{R_f(z+z_1)+vt}{2\sqrt{D_A R_f t}}\right] \tag{7-41}$$

$$\chi(z,t) = \frac{1}{2}\text{erfc}\left[\frac{R_f z - vt}{2\sqrt{D_A R_f t}}\right] + \sqrt{\frac{v^2 t}{\pi D_A R_f}}\exp\left[-\frac{(R_f z - vt)^2}{4D_A R_f t}\right]\ldots$$
$$\ldots - \frac{1}{2}\left(1 + \frac{vz}{D_A} + \frac{v^2 t}{D_A R_f}\right)\exp\left(\frac{vz}{D_A}\right)\text{erfc}\left[\frac{R_f z + vt}{2\sqrt{D_A R_f t}}\right] \tag{7-42}$$

7.3 Numerical Evaluation

Numerical evaluation of the concentration profile requires the calculation of the error function or the complimentary error function. The *error function* is defined as

$$\text{erf}(u) = \frac{2}{\sqrt{\pi}}\int_0^u e^{-\eta^2}\,d\eta \tag{7-43}$$

and the *complimentary error function* as

$$\text{erfc}(u) = 1 - \text{erf}(u) = \frac{2}{\sqrt{\pi}}\int_u^\infty e^{-\eta^2}\,d\eta \tag{7-44}$$

Tabulations of the error function may be found in advanced calculus textbooks and other mathematics handbooks. Error function values can also be obtained from the other sources including Microsoft Excel (with the analysis tool pack add-in), Mathcad, IMSL Fortran Math Library Special Functions (call ERF(X)), Matlab, and other mathematical software packages. Further details about the error function can be found in Appendix A.

7.4 Development

The advection-diffusion equation is solved using a Laplace transform technique. The real partial differential equation in time, t, and space, z, is transformed into a complex ordinary differential equation of space only. This ordinary differential equation was solved with the given transformed boundary conditions.

This solution in the Laplace domain is then inverted back into the time domain with the assistance of Laplace transform inversion tables, such as those found in Carslaw and Jaeger (1959), Özisk (1993), and in Appendix B.

For case 1 the transformed dynamic equation has the form

$$\mathscr{L}\left[R_f \frac{\partial c_A}{\partial t}\right] = \mathscr{L}\left[D_A \frac{\partial^2 c_A}{\partial z^2} - v\frac{\partial c_A}{\partial z}\right]$$
$$R_f s \cdot C(z,s) - R_f c_i = D_A \frac{d^2 c_A}{dz^2} - v\frac{dc_A}{dz} \tag{7-45}$$

where

$$\mathscr{L}\left[c_A(z,t)\right] = C(z,s) \tag{7-46}$$

The boundary conditions that are required to be satisfied are

$$\mathscr{L}\left[c_A(z,t)\right] = \mathscr{L}\left[c_o\right]$$
$$C(z,s) = \frac{c_o}{s} \qquad \text{at } z = 0 \tag{7-47}$$

and the final solution is required to be bounded.

A bounded solution to (7-45) with the above constraint has the form

$$C(z,s) = \frac{c_i}{s} + (c_o - c_i)\frac{1}{s}\mathbf{exp}\left[\frac{vz}{2D_A} - z\sqrt{\frac{v^2}{4D_A{}^2} + \frac{R_f}{D_A}s}\right] \tag{7-48}$$

where the boundary condition can easily be seen as satisfied at $z = 0$, and

$$-v\frac{dC}{dz} = -\left(\frac{c_o - c_i}{s}\right)v\lambda\,\mathbf{exp}\left[\frac{vz}{2D_A} - z\sqrt{\frac{v^2}{4D_1{}^2} + \frac{R_f}{D_A}s}\right] \tag{7-49}$$

$$D_A\frac{d^2C}{dz^2} = \left(\frac{c_o - c_i}{s}\right)D_A\lambda^2\,\mathbf{exp}\left[\frac{vz}{2D_A} - z\sqrt{\frac{v^2}{4D_1{}^2} + \frac{R_f}{D_A}s}\right] \tag{7-50}$$

where

$$\lambda = \frac{v}{2D_A} - \sqrt{\frac{v^2}{4D_A{}^2} + \frac{R_f}{D_A}s} \tag{7-51}$$

As

$$R_f sC(z,s) - R_f c_i = R_f s\left(\frac{c_o - c_i}{s}\right)\exp\left[\frac{vz}{2D_A} - z\sqrt{\frac{v^2}{4D_A{}^2} + \frac{R_f}{D_A}s}\right] \tag{7-52}$$

For (7-52) to equal the sum of equations (7-49) and (7-50), we need to show that

$$D_A\lambda^2 - v\lambda - R_f s = 0 \tag{7-53}$$

The roots to this quadratic function are

$$\lambda = \frac{v \pm \sqrt{v^2 + 4D_A R_f s}}{2D_A} \tag{7-54}$$

As the positive root is equal to (7-51), thus the dynamic equation is satisfied by the solution.

Taking the inverse Laplace transform we obtain the following solution in the time domain:

$$\mathscr{L}^{-1}\left[C(z,s)\right] = c_i \cdot \mathscr{L}^{-1}\left[\frac{1}{s}\right] + (c_o - c_i)\exp\left(\frac{vz}{2D_A}\right)\mathscr{L}^{-1}\left[\frac{1}{s}\exp\left\{-\sqrt{\left(\frac{vz}{2D_A}\right)^2 + \frac{R_f z^2}{D_A}s}\right\}\right]$$

$$c_A(z,t) = c_i +$$

$$(c_o - c_i)\exp\left(\frac{vz}{2D_A}\right)\left[\frac{1}{2}\exp\left(-\frac{vz}{2D_A}\right)\mathrm{erfc}\left\{\frac{R_f z - vt}{\sqrt{4D_A R_f t}}\right\} + \frac{1}{2}\exp\left(\frac{vz}{2D_A}\right)\mathrm{erfc}\left\{\frac{R_f z + vt}{\sqrt{4D_A R_f t}}\right\}\right]$$

$$\therefore\quad c_A(z,t) = c_i + \left(\frac{c_o - c_i}{2}\right)\left[\mathrm{erfc}\left(\frac{R_f z - vt}{\sqrt{4D_A R_f t}}\right) + \exp\left(\frac{vz}{D_A}\right)\mathrm{erfc}\left(\frac{R_f z + vt}{\sqrt{4D_A R_f t}}\right)\right] \tag{7-55}$$

Derivations for the other cases have been omitted. The solution procedures follow the same pattern whereby the integrated form of the ordinary differential equation is required to be known beforehand due to the complexity of the boundary value problem. The forms of the solutions are developed from ordinary differential equation heuristics. Refer to van Genuchten (1981) and Carslaw and Jaeger (1959) for further details of the solution development.

References

van Genuchten, M.Th., Alves, W.J. (1982) *Analytical Solutions of the One-Dimensional Convective-Dispersive Solute Transport Equation*. U.S. Department of Agriculture, Technical Bulletin No. 1661.

van Genuchten, M.Th. (1981) Analytical Solutions for Chemical Transport with Simultaneous Adsorption, Zero-Order Production and First-Order Decay, *J. Hydrology*, **49**, 213-233.

Carslaw, H.S., Jaeger, J.C. (1959) *Conduction of Heat in Solids*, Clarendon Press, Oxford.

Farlow, S.J. (1993) *Partial Differential Equations for Scientists and Engineers*, Dover Publications, New York.

8 Volatile liquid evaporation

8.1 Introduction

The previous models have focused on diffusion or advection and diffusion in a single phase. This following section refers to the modeling of volatile liquid compounds evaporating from subsurface environments. In such a system, the presence of the more dense liquid phase maintains high concentration gradients far longer than would be observed in a homogeneous vapor phase. A basic model involving a pseudo steady-state diffusion from a receding liquid front is used to describe the system behavior. Air emissions from petroleum waste sites have been predicted using such models (Thibodeaux and Hwang, 1982)

A summary of the model with various boundary and initial conditions is given in Section 8.2 with notes regarding their numerical evaluation given in Section 8.3. A detailed development of the model equations is given in Section 8.4.

8.2 Analysis Summary

Assuming the limiting step in the evaporation of a nonaqueous phase liquid from a sediment bed is the downward movement of the vaporizing liquid front. Thus the non-aqueous phase is modeled as pseudo steady-state vapor diffusion through the sediment, from a receding liquid front.

8.2.1 Case 1: Evaporation and vapor diffusion through soil/sediment with uniform initial liquid saturation, with zero vapor concentration at the surface

The evaporation is modeled by vapor diffusion through a sediment column as shown in the following diagram:

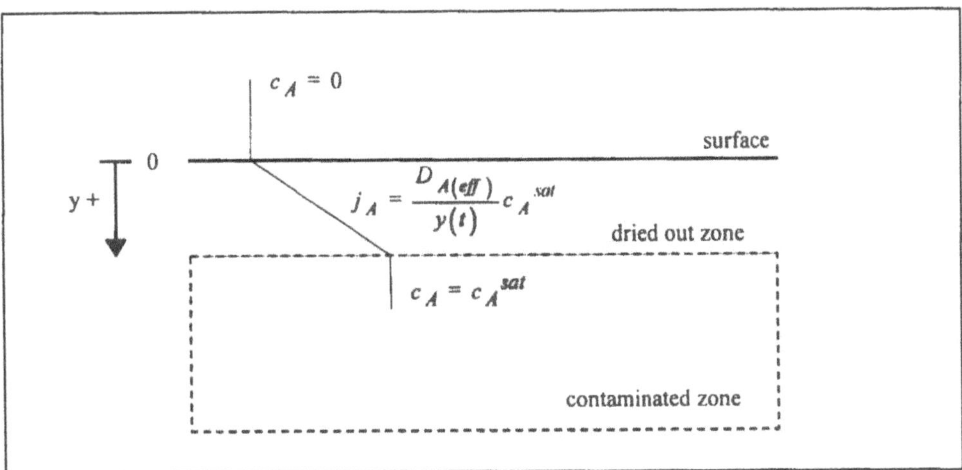

Figure 8-1 Volatile liquid evaporation and vapor diffusion through a sediment column with zero surface concentration.

A pseudo steady-state flux through the depleted or dried out top zone is given by

$$j_A(t) = \frac{D_{A(eff)}}{y(t)}\left[c_A^{sat} - c_A^{\infty}\right] \tag{8-1}$$

Assuming $c_A^{\infty} \to 0$, also the rate in which the depleted zone depth increases is given by

$$\frac{dy}{dt} = \frac{j_A}{\rho_A^{field}} \tag{8-2}$$

where ρ_A^{field} is the mass of volatile organic liquid per unit volume of sediment.

Assuming a uniform initial volatile liquid distribution, the flux rate is given by

$$j_A(t) = \sqrt{\frac{D_{A(eff)} \cdot c_A^{sat} \cdot \rho_A^{field}}{2t}} \tag{8-3}$$

8.2.2 Case 2: Evaporation and vapor diffusion through soil/sediment with uniform initial liquid saturation, with a vapor mass transfer boundary condition at the surface

The evaporation is modeled by vapor diffusion through a sediment column and a convective boundary condition at the surface, as shown in the following diagram:

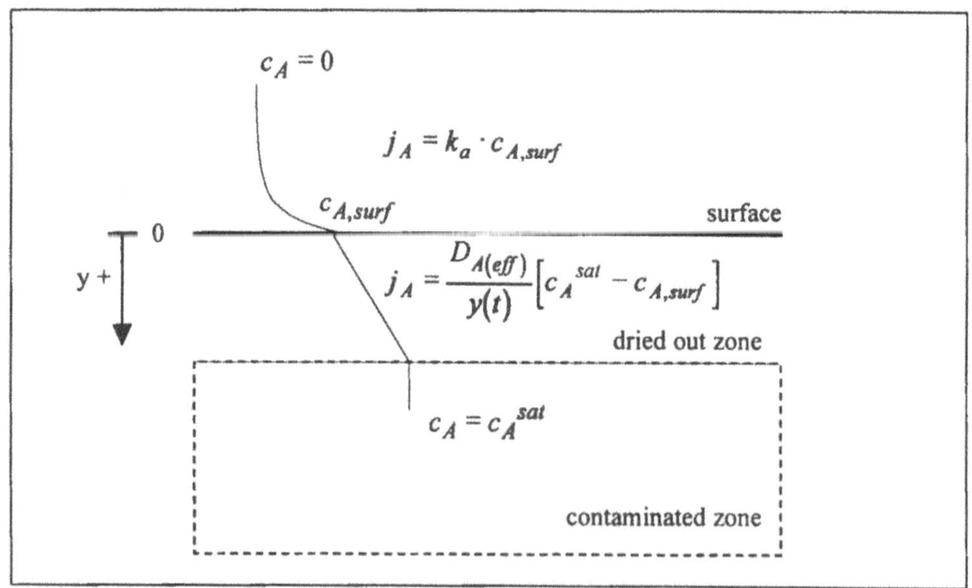

Figure 8-2 Volatile liquid evaporation and vapor diffusion through a sediment column with a mass transfer surface boundary.

A pseudo steady-state flux through the depleted or dried out top zone and convective boundary is given by

$$j_A(t) = \frac{D_{A(eff)}}{y(t)}\left[c_A^{sat} - c_{A,surf}\right] = k_a\left(c_{A,surf} - c_A^{\infty}\right) \tag{8-4}$$

Assuming $c_A^{\infty} \to 0$, also the rate in which the depleted zone depth increases is given by

$$\frac{dy}{dt} = \frac{j_A}{\rho_A^{field}} \tag{8-5}$$

where ρ_A^{field} is the mass of volatile organic liquid per unit volume of sediment.

Assuming a uniform initial volatile liquid distribution, the flux rate is given by

$$j_A(t) = \frac{k_a \cdot c_A^{sat}}{\sqrt{1 + \dfrac{2k_a^2 \cdot c_A^{sat} t}{\rho^{field} \cdot D_{A(eff)}}}} \tag{8-6}$$

8.2.3 Case 3: Evaporation and vapor diffusion through soil/sediment with uniform initial liquid saturation below a finite clean capped region, with zero vapor concentration at the surface

The evaporation is modeled by vapor diffusion through a sediment column as shown in the following diagram:

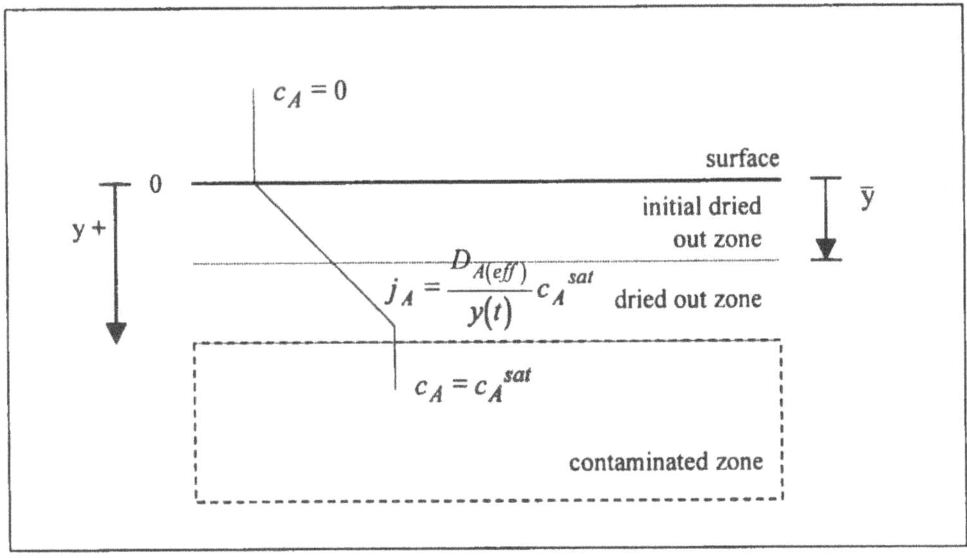

Figure 8-3 Volatile liquid evaporation and vapor diffusion through a sediment column with zero surface concentration and initial depleted cap.

A pseudo steady-state flux through the depleted or dried out top zone is given by

$$j_A(t) = \frac{D_{A(eff)}}{y(t)}\left[c_A^{sat} - c_A^{\infty}\right] \tag{8-7}$$

Assuming $c_A^{\infty} \to 0$, also the rate in which the depleted zone depth increases is given by

$$\frac{dy}{dt} = \frac{j_A}{\rho_A^{field}} \tag{8-8}$$

where ρ_A^{field} is the mass of volatile organic liquid per unit volume of sediment.

Assuming initially no volatile liquid is present above a depth of $y = \bar{y}$ (due to a clean cap or other depletion process) and a uniform volatile liquid distribution of ρ_A^{field} exists below the depth of $y = \bar{y}$, the flux rate is given by

$$j_A(t) = \frac{D_{A(eff)} \cdot c_A^{sat}}{\sqrt{\bar{y}^2 + \dfrac{2D_{A(eff)} \cdot c_A^{sat} \cdot t}{\rho^{field}}}} \tag{8-9}$$

8.2.4 Case 4: Evaporation and vapor diffusion through soil/sediment with uniform initial liquid saturation below a finite clean capped region, with a vapor mass transfer boundary condition at the surface

The evaporation is modeled by vapor diffusion through a sediment column and a convective boundary condition at the surface, as shown in the following diagram:

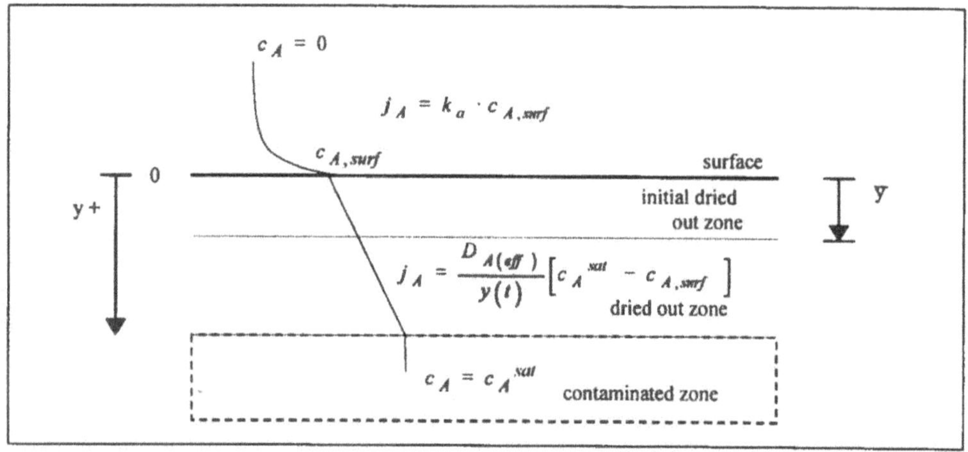

Figure 8-4 Volatile liquid evaporation and vapor diffusion through a sediment column with a mass transfer surface boundary and initial depleted cap.

A pseudo steady-state flux through the depleted or dried out top zone and convective boundary is given by

$$j_A(t) = \frac{D_{A(eff)}}{y(t)}\left[c_A^{sat} - c_{A,surf}\right] = k_a\left(c_{A,surf} - c_A^{\infty}\right) \tag{8-10}$$

Assuming $c_A^{\infty} \to 0$, also the rate in which the depleted zone depth increases is given by

$$\frac{dy}{dt} = \frac{j_A}{\rho_A^{field}} \tag{8-11}$$

where ρ_A^{field} is the mass of volatile organic liquid per unit volume of sediment.

Assuming initially no volatile liquid is present above a depth of $y = \bar{y}$ (due to a clean cap or other depletion process) and a uniform volatile liquid distribution of ρ_A^{field} exists below the depth of $y = \bar{y}$, the flux rate is given by

$$j_A(t) = \frac{k_a \cdot c_A^{sat}}{\sqrt{1 + \dfrac{2k_a \cdot \bar{y}}{D_{A(eff)}} + \left(\dfrac{k_a \cdot \bar{y}}{D_{A(eff)}}\right)^2 + \dfrac{2k_a^2 \cdot c_A^{sat} \cdot t}{\rho^{field} \cdot D_{A(eff)}}}} \tag{8-12}$$

8.3 Numerical Evaluation

No special numerical techniques are required to evaluate the flux rates from the models described in the previous section.

8.4 Development

8.4.1 Case 1: Evaporation and vapor diffusion through soil/sediment with uniform initial liquid saturation, with zero vapor concentration at the surface

Combining equations (8-1) and (8-2) we obtain

$$\frac{dy}{dt} = \frac{D_{A(eff)} \cdot c_A^{sat}}{\rho_A^{field}} \frac{1}{y(t)} \tag{8-13}$$

Solving this with the initial conditions, we obtain

$$\int_0^y y\,dy = \left[\frac{D_{A(eff)} \cdot c_A^{sat}}{\rho_A^{field}}\right]\int_0^t dt \tag{8-14}$$

$$\tfrac{1}{2}y^2 = \left[\frac{D_{A(eff)} \cdot c_A^{sat}}{\rho_A^{field}}\right]t \tag{8-15}$$

Hence the depth of the depleted zone is given by

$$y = \sqrt{\frac{2D_{A(eff)} \cdot c_A^{sat} \cdot t}{\rho_A^{field}}} \tag{8-16}$$

Substituting this into equation (8-1) we obtain a flux rate of

$$j_A(t) = \sqrt{\frac{D_{A(eff)} \cdot c_A^{sat} \cdot \rho_A^{field}}{2t}} \tag{8-17}$$

8.4.2 Case 2: Evaporation and vapor diffusion through soil/sediment with uniform initial liquid saturation, with a vapor mass transfer boundary condition at the surface

Solving equation (8-4) for the surface concentration, we obtain

$$c_{A,surf} = \frac{c_A^{sat}}{1 + \left(\frac{k_a}{D_{A(eff)}} \right) y(t)} \tag{8-18}$$

Hence the flux is given by

$$j_A = \frac{k_a \cdot c_A^{sat}}{1 + \left(\frac{k_a}{D_{A(eff)}} \right) y(t)} \tag{8-19}$$

Combining equations (8-5) and (8-19) we obtain

$$\frac{dy}{dt} = \left(\frac{k_a \cdot c_A^{sat}}{\rho_A^{field}} \right) \frac{1}{1 + \left(\frac{k_A}{D_{A(eff)}} \right) y(t)} \tag{8-20}$$

Solving this with the initial conditions, we obtain

$$\int_0^y \left(1 + \frac{k_a}{D_{A(eff)}} y \right) dy = \left[\frac{k_a \cdot c_A^{sat}}{\rho_A^{field}} \right] \int_0^t dt \tag{8-21}$$

$$y + \frac{k_a}{2D_{A(eff)}} y^2 = \left[\frac{k_a \cdot c_A^{sat}}{\rho_A^{field}} \right] t \tag{8-22}$$

Hence the depth of the depleted zone is given by

$$y(t) = \frac{D_{A(eff)}}{k_a} \left[-1 + \sqrt{1 + \frac{2k_a^2 \cdot c_A^{sat} \cdot t}{\rho^{field} \cdot D_{A(eff)}}} \right] \tag{8-23}$$

Substituting this into equation (8-19) we obtain a flux rate of

$$j_A(t) = \frac{k_a \cdot c_A^{sat}}{\sqrt{1 + \dfrac{2k_a^2 \cdot c_A^{sat} t}{\rho^{field} \cdot D_{A(eff)}}}} \tag{8-24}$$

8.4.3 Case 3: Evaporation and vapor diffusion through soil/sediment with uniform initial liquid saturation below a finite clean capped region, with zero vapor concentration at the surface

Combining equations (8-7) and (8-8) we obtain

$$\frac{dy}{dt} = \frac{D_{A(eff)} \cdot c_A^{sat}}{\rho_A^{field}} \frac{1}{y(t)} \tag{8-25}$$

Solving this with the initial conditions, we obtain

$$\int_{\bar{y}}^{y} y\, dy = \left[\frac{D_{A(eff)} \cdot c_A^{sat}}{\rho_A^{field}} \right] \int_0^t dt \tag{8-26}$$

$$\tfrac{1}{2} y^2 - \tfrac{1}{2} \bar{y}^2 = \left[\frac{D_{A(eff)} \cdot c_A^{sat}}{\rho_A^{field}} \right] t \tag{8-27}$$

Hence the depth of the depleted zone is given by

$$y = \sqrt{\bar{y}^2 + \frac{2D_{A(eff)} \cdot c_A^{sat} \cdot t}{\rho_A^{field}}} \tag{8-28}$$

Substituting this into equation (8-7) we obtain a flux rate of

$$j_A(t) = \frac{D_{A(eff)} \cdot c_A^{sat}}{\sqrt{\bar{y}^2 + \dfrac{2D_{A(eff)} \cdot c_A^{sat} \cdot t}{\rho^{field}}}} \tag{8-29}$$

8.4.4 Case 4: Evaporation and vapor diffusion through soil/sediment with uniform initial liquid saturation below a finite clean capped region, with a vapor mass transfer boundary condition at the surface

Solving equation (8-10) for the surface concentration, we obtain

$$c_{A,surf} = \frac{c_A^{sat}}{1 + \left(\dfrac{k_a}{D_{A(eff)}} \right) y(t)} \tag{8-30}$$

Hence the flux is given by

$$j_A = \frac{k_a \cdot c_A^{sat}}{1 + \left(\frac{k_a}{D_{A(eff)}}\right) y(t)} \tag{8-31}$$

Combining equations (8-11) and (8-31) we obtain

$$\frac{dy}{dt} = \left(\frac{k_a \cdot c_A^{sat}}{\rho_A^{field}}\right) \frac{1}{1 + \left(\frac{k_A}{D_{A(eff)}}\right) y(t)} \tag{8-32}$$

Solving this with the initial conditions, we obtain

$$\int_{\bar{y}}^{y} \left(1 + \frac{k_a}{D_{A(eff)}} y\right) dy = \left[\frac{k_a \cdot c_A^{sat}}{\rho_A^{field}}\right] \int_0^t dt \tag{8-33}$$

$$y + \frac{k_a}{2D_{A(eff)}} y^2 - \bar{y} - \frac{k_a}{2D_{A(eff)}} \bar{y}^2 = \left[\frac{k_a \cdot c_A^{sat}}{\rho_A^{field}}\right] t \tag{8-34}$$

Hence the depth of the depleted zone is given by

$$y(t) = \frac{D_{A(eff)}}{k_a} \left[-1 + \sqrt{1 + \frac{2k_a \cdot \bar{y}}{D_{A(eff)}} + \left(\frac{k_a \cdot \bar{y}}{D_{A(eff)}}\right)^2 + \frac{2k_a^2 \cdot c_A^{sat} \cdot t}{\rho^{field} \cdot D_{A(eff)}}} \right] \tag{8-35}$$

Substituting this into equation (8-19) we obtain a flux rate of

$$j_A(t) = \frac{k_a \cdot c_A^{sat}}{\sqrt{1 + \frac{2k_a \cdot \bar{y}}{D_{A(eff)}} + \left(\frac{k_a \cdot \bar{y}}{D_{A(eff)}}\right)^2 + \frac{2k_a^2 \cdot c_A^{sat} \cdot t}{\rho^{field} \cdot D_{A(eff)}}}} \tag{8-36}$$

References

Thibodeaux, L.J., Hwang, S.T. (1982) Landfarming of Petroleum Wastes – Modeling the Air Emission Problem, *Environ. Progress*, **1**(1), 42-46.

Thibodeaux, L.J. (1996) *Environmental Chemodynamics*, 2nd ed., John Wiley & Sons, New York.

9 Diffusion with time-dependent partition coefficients

9.1 Introduction

In certain situations, in particular in unsaturated soils, time dependent diffusion coefficient models arise. Let us consider vapor transport in an unsaturated soil. The soil-air partition coefficient is defined as the ratio of the equilibrium concentrations as given by

$$K_{soil-air} = \left. \frac{c_A^{soil}}{c_A^{air}} \right|_{equilibrium} \tag{9-1}$$

where the air phase is generally in volumetric concentration [mg (contaminant)/L (air)] and the soil is in mass concentration units [mg (contaminant)/kg (dry soil)].

The value of the partition coefficient is a function of soil and chemical compound properties. Several important soil properties that greatly affect the magnitude of the partition coefficient include the fraction of organic carbon and the moisture content. Further discussion of the equilibrium of contaminants with the soil phase is given in Section 2.1.3.

Moisture content, and thus equilibrium partitioning, of exposed soil often varies with time, e.g., effects due to the diurnal cycle in a dry climate. The following is a mathematical analysis (Section 9.2), which outlines the effect of time-dependent soil-air partition coefficient on the dynamics of a contaminant species.

$$K_{soil-air} = f(t) \tag{9-2}$$

Several cases are then presented, where the soil-air partition coefficient varies with time in a small, finite depth. Expressions for calculating the concentration profiles and surface flux are given in Sections 9.3. Several function forms of the soil-air partition coefficient's time dependence are further analyzed in Section 9.4. The development of the solutions is presented in Section 9.5.

9.2 Mathematical Analysis

As developed in Section 2.4, a microscopic mass balance whereby diffusion in the pore space is the only mechanism of transport results in contaminant behavior described by the following equation:

$$\frac{\partial c_A^{total}}{\partial t} = D_{A(eff)} \frac{\partial^2 c_{A,mobile\ phase}}{\partial z^2} \tag{9-3}$$

where $D_{A(eff)}$ [m^2/s] is the molecular diffusion coefficient of the contaminant species modified for soil's porosity and tortuosity and z [m] and t [s] are the space and time variables. Taking the mobile phase to be the air phase, the total concentration (in the left-hand side, accumulation term) may be separated into its constitute phases and expressed in the basis of the air phase concentration,

$$
\begin{aligned}
c_A^{total} &= \varepsilon \cdot c_A^{air} + (1-\varepsilon)\rho_{soil} \cdot c_A^{soil} \\
&= \varepsilon \cdot c_A^{air} + \rho_b \cdot c_A^{air} K_{soil-air}(t)
\end{aligned} \tag{9-4}
$$

From henceforth using $c_A = c_A^{air}$ and $K_{sa} = K_{soil-air}$ we have

$$
\begin{aligned}
\frac{\partial c_A^{total}}{\partial t} &= \varepsilon \frac{\partial c_A}{\partial t} + \rho_b \frac{\partial}{\partial t}\{c_A(z,t) \cdot K_{sa}(t)\} \\
&= \varepsilon \frac{\partial c_A}{\partial t} + \rho_b K_{sa}(t) \frac{\partial c_A}{\partial t} + \rho_b \frac{dK_{sa}}{dt} c_A(z,t)
\end{aligned} \tag{9-5}
$$

Thus equation (9-3) becomes

$$
\varepsilon \frac{\partial c_A}{\partial t} + \rho_b K_{sa}(t) \frac{\partial c_A}{\partial t} + \rho_b \frac{dK_{sa}}{dt} c_A(z,t) = D_{A(eff)} \frac{\partial^2 c_A}{\partial z^2} \tag{9-6}
$$

$$
\frac{\partial c_A}{\partial t} = \left\{ \frac{D_{A(eff)}}{\varepsilon + \rho_b K_{sa}(t)} \right\} \frac{\partial^2 c_A}{\partial z^2} - \left\{ \frac{\rho_b \frac{dK_{sa}}{dt}}{\varepsilon + \rho_b K_{sa}(t)} \right\} c_A(z,t) \tag{9-7}
$$

In order to solve equation (9-7), two variable transformations are required to remove the time dependence of the equation coefficients. First, let

$$
c_A(z,t) = \phi(z,t) \exp\left[-\int_0^t \frac{\rho_b \frac{dK_{sa}}{dt'}}{\varepsilon + \rho_b K_{sa}(t')} dt' \right] \tag{9-8}
$$

thus giving

$$
\frac{\partial^2 c_A}{\partial z^2} = \frac{\partial^2 \phi}{\partial z^2} \exp\left[-\int_0^t \frac{\rho_b \frac{dK_{sa}}{dt'}}{\varepsilon + \rho_b K_{sa}(t')} dt' \right] \tag{9-9}
$$

$$
\frac{\partial c_A}{\partial t} = \frac{\partial \phi}{\partial t} \exp\left[-\int_0^t \frac{\rho_b \frac{dK_{sa}}{dt'}}{\varepsilon + \rho_b K_{sa}(t')} dt' \right] - \left\{ \frac{\rho_b \frac{dK_{sa}}{dt}}{\varepsilon + \rho_b K_{sa}(t)} \right\} \phi(z,t) \exp\left[-\int_0^t \frac{\rho_b \frac{dK_{sa}}{dt'}}{\varepsilon + \rho_b K_{sa}(t')} dt' \right] \tag{9-10}
$$

Substituting equations (9-8) to (9-10) into equation (9-7), and dividing by the common exponential factor we obtain

$$
\frac{\partial \phi}{\partial t} - \left\{ \frac{\rho_b \frac{dK_{sa}}{dt}}{\varepsilon + \rho_b K_{sa}(t)} \right\} \phi(z,t) = \left\{ \frac{D_{A(eff)}}{\varepsilon + \rho_b K_{sa}(t)} \right\} \frac{\partial^2 \phi}{\partial z^2} - \left\{ \frac{\rho_b \frac{dK_{sa}}{dt}}{\varepsilon + \rho_b K_{sa}(t)} \right\} \phi(z,t)
$$

$$\therefore \quad \frac{\partial \phi}{\partial t} = \left\{ \frac{D_{A(eff)}}{\varepsilon + \rho_b K_{sa}(t)} \right\} \frac{\partial^2 \phi}{\partial z^2} \tag{9-11}$$

Second, let

$$\tau = \int_0^t \frac{D_{A(eff)}}{\varepsilon + \rho_b K_{sa}(t')} dt' \tag{9-12}$$

hence giving

$$\frac{d\tau}{dt} = \frac{D_{A(eff)}}{\varepsilon + \rho_b K_{sa}(t)} \tag{9-13}$$

Transforming the time variable in equation (9-11) to $\tau(t)$ we obtain

$$\begin{aligned} \frac{\partial \phi}{\partial \tau} &= \frac{\partial \phi}{\partial t} \cdot \frac{dt}{d\tau} \\ &= \left\{ \frac{D_{A(eff)}}{\varepsilon + \rho_b K_{sa}(t)} \right\} \frac{\partial^2 \phi}{\partial z^2} \left\{ \frac{\varepsilon + \rho_b K_{sa}(t)}{D_{A(eff)}} \right\} \end{aligned} \tag{9-14}$$

$$\therefore \quad \frac{\partial \phi}{\partial \tau} = \frac{\partial^2 \phi}{\partial z^2} \tag{9-15}$$

Equation (9-15) can be solved for $\phi(z,t)$ using standard methods (examples are shown in Section 9.5) for z and t, where t has been transformed into $\tau(t)$ by equation (9-12). Concentration $c_A(z,t)$ can then be obtained from $\phi(z,t)$ by equation (9-8).

9.3 Analysis Summary

9.3.1 Case 1: Diffusion in a thin layer with time-dependent soil-air partition coefficient, zero surface concentration, a no-flow bottom boundary condition, and constant initial conditions

For the system defined by the following dynamics and boundary conditions

$$\frac{\partial}{\partial t}\{\varepsilon \cdot c_A(z,t) + \rho_b \cdot c_A(z,t) \cdot K_{sa}(t)\} = D_{A(eff)} \frac{\partial^2 c_A}{\partial z^2} \qquad z = [0, L] \tag{9-16}$$

$$c_A(z,t)\big|_{z=L} = 0 \qquad t > 0 \tag{9-17}$$

$$\left.\frac{\partial c_A}{\partial z}\right|_{z=0} = 0 \qquad t > 0 \tag{9-18}$$

$$c_A(z,t)\big|_{t=0} = c_{A0} \qquad z = [0, L] \tag{9-19}$$

This system is illustrated by the following diagram:

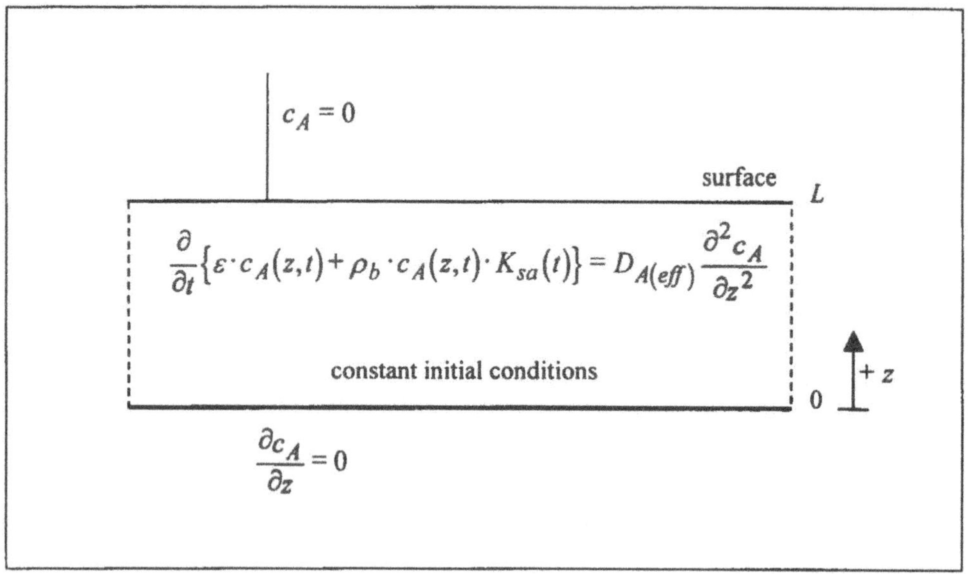

Figure 9-1 Diffusion in a thin layer with time-dependent soil-air partition coefficient, zero surface concentration, a no-flow bottom boundary condition, and constant initial conditions.

The mobile phase concentration profile is given by

$$c_A(z,t) = -\frac{2c_{A0}}{L} \exp\left[-\int_0^t \frac{\rho_b \frac{dK_{sa}}{dt'}}{\varepsilon + \rho_b K_{sa}(t')} dt'\right] \sum_{n=1}^{\infty} \exp\left\{-\beta_n^2 \tau(t)\right\} \frac{(-1)^n}{\beta_n} \cos(\beta_n z) \quad (9\text{-}20)$$

and the surface flux is given by

$$j_A(t) = \frac{2D_{A(eff)} c_{A0}}{L} \exp\left[-\int_0^t \frac{\rho_b \frac{dK_{sa}}{dt'}}{\varepsilon + \rho_b K_{sa}(t')} dt'\right] \sum_{n=1}^{\infty} \exp\left\{-\beta_n^2 \tau(t)\right\} \quad (9\text{-}21)$$

where

$$\beta_n = \frac{\pi}{L}\left(n - \frac{1}{2}\right) \qquad n = 1,2,3... \quad (9\text{-}22)$$

and

$$\tau = \int_0^t \frac{D_{A(eff)}}{\varepsilon + \rho_b K_{sa}(t')} dt' \quad (9\text{-}23)$$

9.3.1.1 *Sample application of case 1*

Evaluation of the general solution is very much influenced by the functional form of the air-soil partition coefficient with respect to time. The following is a specific example of an application of case 1's general solution. This is presented to demonstrate the numerical procedure required for a given case of $K_{sa} = f(t)$. The example was used to model experiments performed by Valsaraj et al. (1997).

Consider a thin layer of sediment within an experimental microcosm. Initially damp, contaminated sediment is contained in the microcosm. Dry air is passed over the top of the contaminated sediment. As this sediment dries over a period of time, the air-soil partition coefficient is expected to have the following approximate behavior:

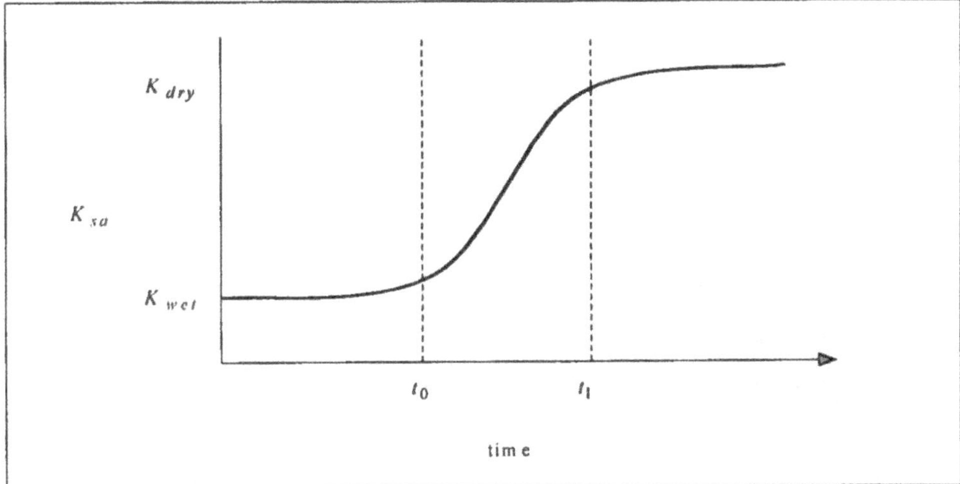

Figure 9-2 Typical time vs. soil-air partition coefficient profile for drying sediment.

Initially the partition coefficient is predicted for wet sediment equilibrium. As dry air is passed over the sediment, the moisture content begins to fall. At the beginning, the partition coefficient is not influenced greatly by this loss in moisture, until the stage where less than a monolayer of water is available to cover the sediment particles. As more of the dry sediment surface is exposed, the contaminant species will have more sites for adsorption. This leads to an increase in the equilibrium partitioning toward the soil phase. Eventually the soil becomes dry and the partition coefficient levels off to a fixed value. Further details about air-water-soil partitioning can be found in Section 2.1.3.

Mathematically, K_{sa}'s behavior may be expressed as

$$K_{sa}(t) = \begin{cases} K_{wet} & \text{for } 0 \le t \le t_0 \\ \left(\dfrac{K_{wet}t_1 - K_{dry}t_0}{t_1 - t_0}\right) + \left(\dfrac{K_{dry} - K_{wet}}{t_1 - t_0}\right)t & \text{for } t_0 \le t \le t_1 \\ K_{dry} & \text{for } t \ge t_1 \end{cases} \tag{9-24}$$

Calculating the time transformation given by equation (9-12) for each time period we obtain

$$\tau = \frac{D_{A(eff)}t}{\varepsilon + \rho_b K_{wet}} \qquad t \in [0, t_0] \quad (9\text{-}25)$$

$$\tau = \frac{D_{A(eff)}t_0}{\varepsilon + \rho_b K_{wet}} + \frac{D_{A(eff)}(t_1 - t_0)}{\rho_b (K_{dry} - K_{wet})} \ln\left\{\frac{A + t}{A + t_0}\right\} \qquad t \in [t_0, t_1] \quad (9\text{-}26)$$

$$\tau = \frac{D_{A(eff)}t_0}{\varepsilon + \rho_b K_{wet}} + \frac{D_{A(eff)}(t_1 - t_0)}{\rho_b (K_{dry} - K_{wet})} \ln\left\{\frac{A + t_1}{A + t_0}\right\} + \frac{D_{A(eff)}(t - t_1)}{\varepsilon + \rho_b K_{dry}} \qquad t \in [t_1, \infty) \quad (9\text{-}27)$$

where

$$A = \frac{1}{\rho_b}\left[\varepsilon + \rho_b\left(\frac{K_{wet}t_1 - K_{dry}t_0}{t_1 - t_0}\right)\right]\left(\frac{t_1 - t_0}{K_{dry} - K_{wet}}\right) \quad (9\text{-}28)$$

Calculating the exponent that contains the integral expression in equation (9-8) for each time period, we obtain

$$\exp\left[-\int_0^t \frac{\rho_b \frac{dK_{sa}}{dt'}}{\varepsilon + \rho_b K_{sa}(t')} dt'\right] = 1 \qquad t \in [0, t_0] \quad (9\text{-}29)$$

$$\exp\left[-\int_0^t \frac{\rho_b \frac{dK_{sa}}{dt'}}{\varepsilon + \rho_b K_{sa}(t')} dt'\right] = \left\{\frac{A + t_0}{A + t}\right\} \qquad t \in [t_0, t_1] \quad (9\text{-}30)$$

$$\exp\left[-\int_0^t \frac{\rho_b \frac{dK_{sa}}{dt'}}{\varepsilon + \rho_b K_{sa}(t')} dt'\right] = \left\{\frac{A + t_0}{A + t_1}\right\} \qquad t \in [t_1, \infty) \quad (9\text{-}31)$$

with A being defined as given in equation (9-28).

Thus the flux is given by

$$\dot{n}_A(t) = \begin{cases} \dfrac{2D_{A(eff)}c_{A0}}{L} \displaystyle\sum_{n=1}^{\infty} \exp\left\{-\beta_n^2 \tau(t)\right\} & \text{for } 0 \le t \le t_0 \\ \dfrac{2D_{A(eff)}c_{A0}}{L} \left\{\dfrac{A+t_0}{A+t}\right\} \displaystyle\sum_{n=1}^{\infty} \exp\left\{-\beta_n^2 \tau(t)\right\} & \text{for } t_0 \le t \le t_1 \\ \dfrac{2D_{A(eff)}c_{A0}}{L} \left\{\dfrac{A+t_0}{A+t_1}\right\} \displaystyle\sum_{n=1}^{\infty} \exp\left\{-\beta_n^2 \tau(t)\right\} & \text{for } t \ge t_1 \end{cases} \qquad (9\text{-}32)$$

where $\tau(t)$ are defined in equations (9-25) to (9-28) and A is defined in equation (9-28).

9.3.2 Case 2: Diffusion time-dependent partition coefficient, zero surface concentration, no flow bottom boundary, and arbitrary initial conditions

For the system defined by the following dynamics and boundary conditions

$$\frac{\partial}{\partial t}\left\{\varepsilon \cdot c_A(z,t) + \rho_b \cdot c_A(z,t) \cdot K_{sa}(t)\right\} = D_{A(eff)} \frac{\partial^2 c_A}{\partial z^2} \qquad z = [0, L] \qquad (9\text{-}33)$$

$$c_A(z,t)\big|_{z=L} = 0 \qquad t > 0 \qquad (9\text{-}34)$$

$$\left.\frac{\partial c_A}{\partial z}\right|_{z=0} = 0 \qquad t > 0 \qquad (9\text{-}35)$$

$$c_A(z,t)\big|_{t=0} = c_{A0}(z) \qquad z = [0, L] \qquad (9\text{-}36)$$

This system is illustrated by the following diagram:

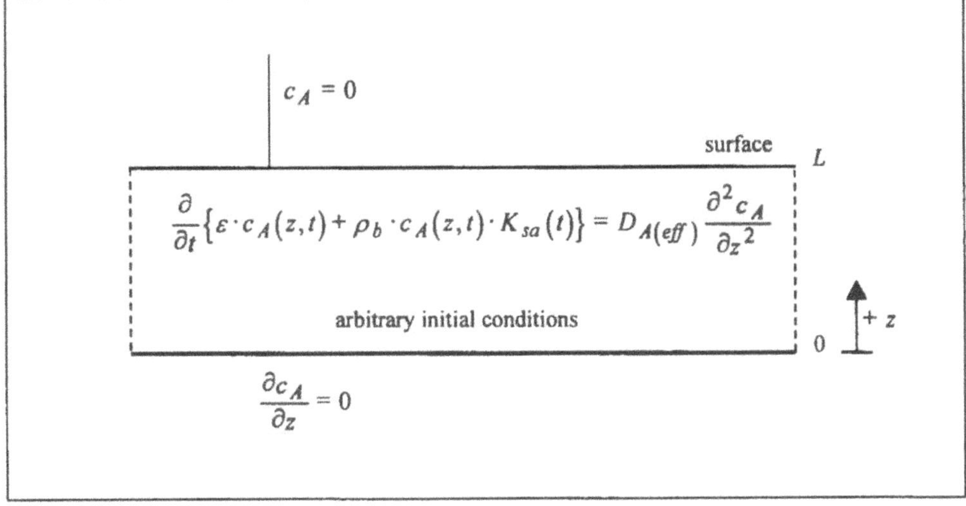

Figure 9-3 Diffusion with time-dependent partition coefficient, zero surface concentration, a no-flow bottom boundary, and arbitrary initial conditions.

The mobile phase concentration profile is given by

$$c_A(z,t)=\frac{2}{L}\exp\left[-\int_0^t\frac{\rho_b\frac{dK_{sa}}{dt'}}{\varepsilon+\rho_bK_{sa}(t')}dt'\right]\sum_{n=1}^{\infty}\exp\left\{-\beta_n{}^2\tau(t)\right\}\cos(\beta_nz)\int_0^L\cos(\beta_nz')c_{A0}(z')dz' \tag{9-37}$$

and the surface flux is given by

$$j_A(t)=-\frac{2D_{A(eff)}}{L}\exp\left[-\int_0^t\frac{\rho_b\frac{dK_{sa}}{dt'}}{\varepsilon+\rho_bK_{sa}(t')}dt'\right]\sum_{n=1}^{\infty}\exp\left\{-\beta_n{}^2\tau(t)\right\}(-1)^n\beta_n\int_0^L\cos(\beta_nz')c_{A0}(z')dz' \tag{9-38}$$

where

$$\beta_n=\frac{\pi}{L}\left(n-\tfrac{1}{2}\right) \qquad n=1,2,3... \tag{9-39}$$

and

$$\tau=\int_0^t\frac{D_{A(eff)}}{\varepsilon+\rho_bK_{sa}(t')}dt' \tag{9-40}$$

9.3.3 Case 3: Diffusion in a thin surface boundary layer with time-dependent soil-air partition coefficient, zero surface concentration, a constant concentration source at the lower boundary, and constant initial conditions

For the system defined by thc following dynamics and boundary conditions

$$\frac{\partial}{\partial t}\left\{\varepsilon\cdot c_A(z,t)+\rho_b\cdot c_A(z,t)\cdot K_{sa}(t)\right\}=D_{A(eff)}\frac{\partial^2c_A}{\partial z^2} \qquad z=[0,L] \tag{9-41}$$

$$c_A(z,t)\Big|_{z=L}=0 \qquad t>0 \tag{9-42}$$

$$c_A(z,t)\Big|_{z=0}=c_{A0} \qquad t>0 \tag{9-43}$$

$$c_A(z,t)\Big|_{t=0}=c_{A0} \qquad z=[0,L] \tag{9-44}$$

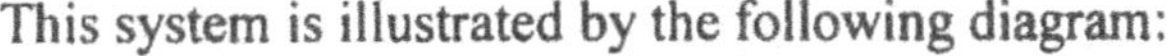

This system is illustrated by the following diagram:

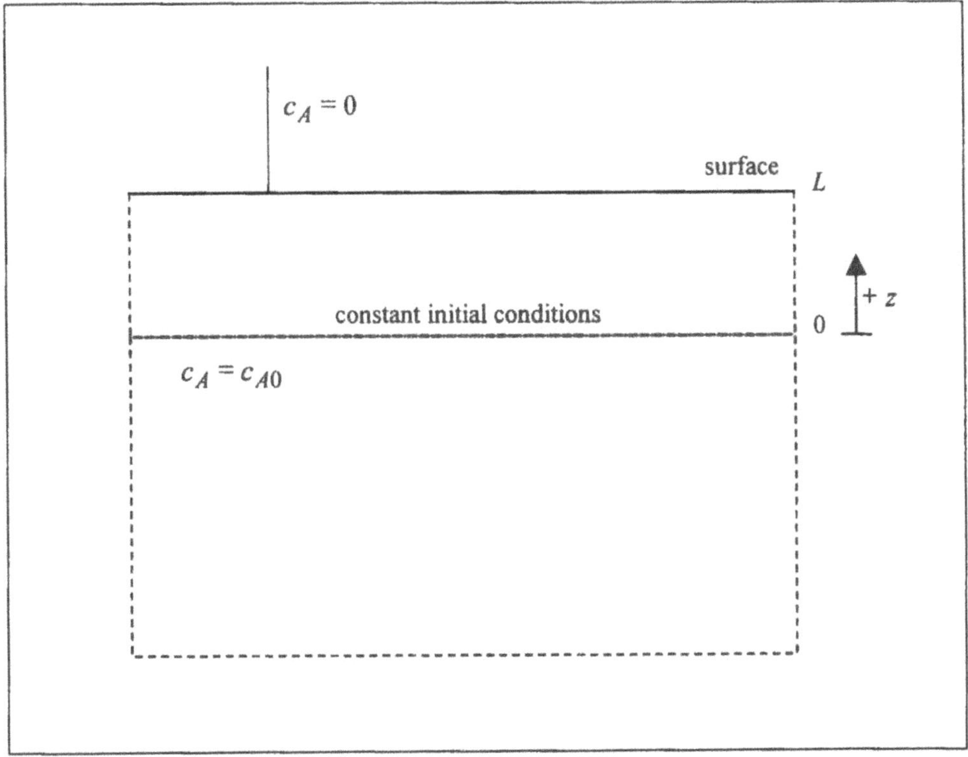

Figure 9-4 Diffusion in a thin surface boundary layer with time-dependent soil-air partition coefficient, zero surface concentration, a constant concentration source at the lower boundary, and constant initial conditions.

The mobile phase concentration profile is given by

$$c_A(z,t) = u(z,t) \cdot c_{A0} \exp\left[-\int_0^t \frac{\rho_b \frac{dK_{sa}}{dt}}{\varepsilon + \rho_b K_{sa}(t)} dt \right] \tag{9-45}$$

where

$$u(z,t) = u_1(z,t) + u_2(z,t) \tag{9-46}$$

$$u_1(z,t) = \left(1 - \frac{z}{L}\right) - \frac{2}{\pi} \sum_{n=1}^{\infty} \frac{(-1)^n}{n} \sin\left(\frac{n\pi}{L} z\right) \exp\left\{ -\left(\frac{n\pi}{L}\right)^2 \int_0^t \frac{D_{A(eff)}}{\varepsilon + \rho_b K_{sa}(t)} dt \right\} \tag{9-47}$$

$$u_2(z,t)=\int_{t'=0}^{t}\left\{\begin{array}{l}\left[\left(1-\frac{z}{L}\right)-\frac{2}{\pi}\sum_{n=1}^{\infty}\frac{1}{n}\sin\left(\frac{n\pi}{L}z\right)\exp\left\{-\left(\frac{n\pi}{L}\right)^2\int_{t'}^{t}\frac{D_{A(eff)}}{\varepsilon+\rho_b K_{sa}(t)}dt\right\}\right]\\ *\left(\frac{\rho_b\frac{dK_{sa}}{dt'}}{\varepsilon+\rho_b K_{sa}(t')}\right)\exp\left[\int_0^{t'}\frac{\rho_b\frac{dK_{sa}}{dt}}{\varepsilon+\rho_b K_{sa}(t)}dt\right]\end{array}\right\}dt' \quad (9\text{-}48)$$

and the surface flux is given by

$$j_A(t)=-D_{A(eff)}c_{A0}\exp\left[-\int_0^{t}\frac{\rho_b\frac{dK_{sa}}{dt}}{\varepsilon+\rho_b K_{sa}(t)}dt\right]\left.\frac{\partial u}{\partial z}\right|_{z=L} \quad (9\text{-}49)$$

where

$$\left.\frac{\partial u}{\partial z}\right|_{z=L}=\left.\frac{\partial u_1}{\partial z}\right|_{z=L}+\left.\frac{\partial u_2}{\partial z}\right|_{z=L} \quad (9\text{-}50)$$

$$\left.\frac{\partial u_1}{\partial z}\right|_{z=L}=-\frac{1}{L}-\frac{2}{L}\sum_{n=1}^{\infty}\exp\left\{-\left(\frac{n\pi}{L}\right)^2\int_0^{t}\frac{D_{A(eff)}}{\varepsilon+\rho_b K_{sa}(t)}dt\right\} \quad (9\text{-}51)$$

$$\frac{\partial u_2}{\partial z}=\frac{1}{L}\int_{t'=0}^{t}\left\{\begin{array}{l}\left[-1-2\sum_{n=1}^{\infty}\exp\left\{-\left(\frac{n\pi}{L}\right)^2\int_{t'}^{t}\frac{D_{A(eff)}}{\varepsilon+\rho_b K_{sa}(t)}dt\right\}\right]\\ *\left(\frac{\rho_b\frac{dK_{sa}}{dt'}}{\varepsilon+\rho_b K_{sa}(t')}\right)\exp\left[\int_0^{t'}\frac{\rho_b\frac{dK_{sa}}{dt}}{\varepsilon+\rho_b K_{sa}(t)}dt\right]\end{array}\right\}dt' \quad (9\text{-}52)$$

9.3.3.1 Sample application for case 3

Evaluation of the general solution is very much influenced by the functional form of the air-soil partition coefficient with respect to time. The following is a specific example of an application of case 3's general solution. This is presented to demonstrate the numerical procedure required for a given case of $K_{sa}=f(t)$.

This example uses the same air-soil partition coefficient vs. time profile as given in Figure 9-2, which represents typical sediment drying behavior. Mathematically, K_d's behavior may be expressed as

$$K_{sa}(t)=\begin{cases} K_{wet} & \text{for } 0\le t\le t_0 \\ \left(\dfrac{K_{wet}t_1-K_{dry}t_0}{t_1-t_0}\right)+\left(\dfrac{K_{dry}-K_{wet}}{t_1-t_0}\right)t & \text{for } t_0\le t\le t_1 \\ K_{dry} & \text{for } t\ge t_1 \end{cases} \tag{9-53}$$

Let us consider each region of time separately. For $t\in[0,t_0]$, equation (9-8) simply becomes

$$c_A(z,t)=u(z,t)\cdot c_{A0} \tag{9-54}$$

and equation (9-48) becomes

$$u_2(z,t)=0 \tag{9-55}$$

as $\dfrac{dK_{sa}}{dt}=0$ in this region of time.

Calculating equation (9-47) explicitly we obtain

$$u_1(z,t)=\left(1-\frac{z}{L}\right)-\frac{2}{\pi}\sum_{n=1}^{\infty}\frac{(-1)^n}{n}\sin\left(\frac{n\pi}{L}z\right)\exp\left[-\left(\frac{n\pi}{L}\right)^2\frac{D_{A(eff)}t}{\varepsilon+\rho_b K_{wet}}\right] \tag{9-56}$$

For the time region $t\in[t_0,t_1]$ equation (9-8) becomes

$$c_A(z,t)=u(z,t)\cdot c_{A0}\left\{\frac{A+t_0}{A+t}\right\} \tag{9-57}$$

and equation (9-47) becomes

$$u_1(z,t)=\left(1-\frac{z}{L}\right)-\frac{2}{\pi}\sum_{n=1}^{\infty}\left\{\begin{array}{l}\dfrac{(-1)^n}{n}\sin\left(\dfrac{n\pi}{L}z\right)\\ *\exp\left[-\left(\dfrac{n\pi}{L}\right)^2\left\{\dfrac{D_{A(eff)}t_0}{\varepsilon+\rho_b K_{wet}}+\dfrac{D_{A(eff)}(t_1-t_0)}{\rho_b(K_{dry}-K_{wet})}\ln\left(\dfrac{A+t}{A+t_0}\right)\right\}\right]\end{array}\right\} \tag{9-58}$$

where

$$A=\frac{1}{\rho_b}\left[\varepsilon+\rho_b\left(\frac{K_{wet}t_1-K_{dry}t_0}{t_1-t_0}\right)\right]\left(\frac{t_1-t_0}{K_{dry}-K_{wet}}\right) \tag{9-59}$$

Equation (9-48) is integrated piecewise by the following:

$$u_2(z,t)=\int_{t'=0}^{t_0}\{\ldots\}dt'+\int_{t'=t_0}^{t}\{\ldots\}dt' \tag{9-60}$$

The first of these integrals is equal to zero as $\frac{dK_{sa}}{dt} = 0$ in this region of time. The second integral is evaluated as

$$u_2(z,t) = \int_{t'=t_0}^{t} \left\{ \begin{array}{l} \left[\left[\left(1-\frac{z}{L}\right) - \frac{2}{\pi}\sum_{n=1}^{\infty}\frac{1}{n}\sin\left(\frac{n\pi}{L}z\right)\exp\left[-\left(\frac{n\pi}{L}\right)^2 \int_{t'}^{t}\frac{D_{A(eff)}}{\varepsilon+\rho_b K_{sa}(t)}dt\right]\right]\right] \\ * \left[\frac{\rho_b \frac{dK_{sa}}{dt'}}{\varepsilon+\rho_b K_{sa}(t')}\exp\left[\int_0^{t'}\frac{\rho_b \frac{dK_{sa}}{dt}}{\varepsilon+\rho_b K_{sa}(t)}dt\right]\right] \end{array} \right\} dt'$$

(9-61)

where

4r $$\int_{t'}^{t}\frac{D_{A(eff)}}{\varepsilon+\rho_b K_{sa}(t)}dt = \frac{D_{A(eff)}(t_1-t_0)}{\rho_b(K_{dry}-K_{wet})}\ln\left\{\frac{A+t}{A+t'}\right\} \tag{9-62}$$

and

$$\begin{aligned} &\frac{\rho_b \frac{dK_{sa}}{dt'}}{\varepsilon+\rho_b K_{sa}(t')}\exp\left[\int_0^{t'}\frac{\rho_b \frac{dK_{sa}}{dt}}{\varepsilon+\rho_b K_{sa}(t)}dt\right] \\ &= \frac{\rho_b \frac{dK_{sa}}{dt'}}{\varepsilon+\rho_b K_{sa}(t')}\exp\left[\int_0^{t_0}\frac{\rho_b \frac{dK_{sa}}{dt}}{\varepsilon+\rho_b K_{sa}(t)}dt + \int_{t_0}^{t'}\frac{\rho_b \frac{dK_{sa}}{dt}}{\varepsilon+\rho_b K_{sa}(t)}dt\right] \\ &= \frac{\rho_b \frac{dK_{sa}}{dt'}}{\varepsilon+\rho_b K_{sa}(t')}\exp\left[\int_{t_0}^{t'}\frac{\rho_b \frac{dK_{sa}}{dt}}{\varepsilon+\rho_b K_{sa}(t)}dt\right] \\ &= \frac{1}{A+t_0} \end{aligned} \tag{9-63}$$

Thus performing the integration, we obtain

$$u_2(z,t) = \left\{\frac{1}{A+t_0}\right\}\left\{\left(1-\frac{z}{L}\right)\int_{t'=t_0}^{t}dt' - \frac{2}{\pi}\sum_{n=1}^{\infty}\frac{1}{n}\sin\left(\frac{n\pi}{L}z\right)\int_{t'=t_0}^{t}\left\{\frac{A+t}{A+t'}\right\}^{-\lambda(n)}dt'\right\}$$

(9-64)

where $\lambda(n)=\left(\frac{n\pi}{L}\right)^2\left\{\frac{D_{A(eff)}(t_1-t_0)}{\rho_b\left(K_{dry}-K_{wet}\right)}\right\}$. Thus

$$u_2(z,t)=$$

$$\left\{\frac{1}{A+t_0}\right\}\left\{\begin{array}{l}\left(1-\frac{z}{L}\right)(t-t_0)\\ -\frac{2}{\pi}\sum_{n=1}^{\infty}\frac{1}{n}\sin\left(\frac{n\pi}{L}z\right)\left\{\left(\frac{A+t}{\lambda(n)+1}\right)-\left(\frac{A+t_0}{\lambda(n)+1}\right)\left(\frac{A+t}{A+t_0}\right)^{-\lambda(n)}\right\}\end{array}\right\} \tag{9-65}$$

For the time region $t\in[t_1,\infty)$ equation (9-8) becomes

$$c_A(z,t)=u(z,t)\cdot c_{A0}\left\{\frac{A+t_0}{A+t_1}\right\} \tag{9-66}$$

and equation (9-47) becomes

$$u_1(z,t)=\left(1-\frac{z}{L}\right)-\frac{2}{\pi}\sum_{n=1}^{\infty}\left\{\begin{array}{l}\frac{(-1)^n}{n}\sin\left(\frac{n\pi}{L}z\right)\\ *\exp\left[-\left(\frac{n\pi}{L}\right)^2\left\{\begin{array}{l}\frac{D_{A(eff)}t_0}{\varepsilon+\rho_bK_{wet}}+\frac{D_{A(eff)}(t_1-t_0)}{\rho_b\left(K_{dry}-K_{wet}\right)}\ln\left(\frac{A+t_1}{A+t_0}\right)\\ +\frac{D_{A(eff)}(t-t_1)}{\varepsilon+\rho_bK_{wet}}\end{array}\right\}\right]\end{array}\right\} \tag{9-67}$$

Equation (9-48) is integrated piecewise by the following:

$$u_2(z,t)=\int_{t'=0}^{t_0}\{\ldots\}dt'+\int_{t'=t_0}^{t_1}\{\ldots\}dt'+\int_{t'=t_1}^{t}\{\ldots\}dt' \tag{9-68}$$

The first and last of these integrals are equal to zero as $\frac{dK_{sa}}{dt}=0$ in these regions of time. The second integral is evaluated as

$$u_2(z,t)=$$

$$\left\{\frac{1}{A+t_0}\right\}\left\{\begin{array}{l}\left(1-\frac{z}{L}\right)(t_1-t_0)\\ -\frac{2}{\pi}\sum_{n=1}\frac{1}{n}\sin\left(\frac{n\pi}{L}z\right)\left\{\left(\frac{A+t_1}{\lambda(n)+1}\right)-\left(\frac{A+t_0}{\lambda(n)+1}\right)\left(\frac{A+t_1}{A+t_0}\right)^{\lambda(n)}\right\}\end{array}\right\} \tag{9-69}$$

It is interesting to note that the above expression is independent of the time variable.

9.4 Variable transformations on a variety of time-dependent air-soil partition coefficient functions

The analysis of contaminant dynamics, with time-dependent soil-air partition coefficients, requires the evaluation of two integrals involving $K_{sa} = f(t)$. These two transformations are given by equations (9-8) and (9-12).

Following are the transformations calculated explicitly for three simple functional forms: $K_{sa}(t)$ is constant, linear, and exponential. The intergrals are calculated over an arbitrary time period $t \in [t_1, t_2]$ as $K_{sa}(t)$ may be made up piecewise from a number of functional sections; the integrals are given for the arbitrary time intervals. For example, if

$$K_{sa}(t) = \begin{cases} constant & \text{for } t \in [0, t_a] \\ linear & \text{for } t \in [t_a, t_b] \\ constant & \text{for } t \in [t_b, \infty) \end{cases} \tag{9-70}$$

then the transformation given by equation (9-12) may be calculated piecewise by

$$\tau = \int_0^{t_a} \frac{D_{A(eff)}}{\varepsilon + \rho_b K_{sa}(t')} dt' \Bigg|_{constant} + \int_{t_a}^{t_b} \frac{D_{A(eff)}}{\varepsilon + \rho_b K_{sa}(t')} dt' \Bigg|_{linear} + \int_{t_b}^{t} \frac{D_{A(eff)}}{\varepsilon + \rho_b K_{sa}(t')} dt' \Bigg|_{constant} \tag{9-71}$$

9.4.1 Constant soil-air partition coefficient

For the case where the soil-air partition coefficient is modeled by the following constant

$$K_{sa}(t) = K_{sa} \tag{9-72}$$

Thus

$$\frac{dK_{sa}}{dt} = 0 \tag{9-73}$$

Equation (9-8) simplifies to

$$c_A(z,t) = \phi(z,t) \tag{9-74}$$

and equation (9-12) simplifies to

$$\tau = \frac{D_{A(eff)}(t_2 - t_1)}{\varepsilon + \rho_b K_{sa}} \tag{9-75}$$

where the integrals were taken from $t \in [t_1, t_2]$.

9.4.2 Linear soil-air partition coefficient

For the case where the soil-air partition coefficient is modeled by the following linear expression

$$K_{sa}(t) = m \cdot t + c \tag{9-76}$$

Thus

$$\frac{dK_{sa}}{dt} = m \tag{9-77}$$

Equation (9-8) simplifies to

$$c_A(z,t) = \phi(z,t) \cdot \left\{ \frac{A + t_1}{A + t_2} \right\} \tag{9-78}$$

and equation (9-12) simplifies to

$$\tau = \frac{D_{A(eff)}}{\rho_b m} \ln \left\{ \frac{A + t_2}{A + t_1} \right\} \tag{9-79}$$

where the integrals were taken from $t \in [t_1, t_2]$ and $A = \left(\frac{\varepsilon + \rho_b c}{\rho_b m} \right)$

9.4.3 Exponential soil-air partition coefficient

For the case where the partition coefficient is modeled by the following exponential function

$$K_{sa}(t) = K_0 \exp(\kappa t) \tag{9-80}$$

Thus

$$\frac{dK_{sa}}{dt} = K_0 \kappa \cdot \exp(\kappa t) \tag{9-81}$$

Equation (9-8) simplifies to

$$c_A(z,t) = \phi(z,t) \left\{ \frac{B + \exp(-\kappa t_1)}{B + \exp(-\kappa t_2)} \right\} \exp[-\kappa (t_2 - t_1)] \tag{9-82}$$

and equation (9-12) simplifies to

$$\tau = \frac{D_{A(eff)}}{\varepsilon \kappa} \left[\ln \left\{ \frac{(1/B) + \exp(\kappa t_1)}{(1/B) + \exp(\kappa t_2)} \right\} + \kappa (t_2 - t_1) \right] \tag{9-83}$$

where the integrals were taken from $t \in [t_1, t_2]$ and $B = \left(\frac{\rho_b K_0}{\varepsilon} \right)$

9.5 Development

9.5.1 Transformation of variables

In all the cases, the dynamic equation is transformed to

$$\frac{\partial \phi}{\partial \tau}=\frac{\partial^2 \phi}{\partial z^2} \tag{9-84}$$

by using equations (9-8) and (9-12). Applying these equations to the boundary conditions, we obtain the following for case 1:

$$\phi(z,\tau)\big|_{z=L}=0 \qquad \tau>0 \tag{9-85}$$

$$\left.\frac{\partial \phi}{\partial z}\right|_{z=0}=0 \qquad \tau>0 \tag{9-86}$$

$$\phi(z,\tau)\big|_{\tau=0}=c_{A0} \qquad z=[0,L] \tag{9-87}$$

Similarly for case 2, except equation (9-87) is replaced by

$$\phi(z,\tau)\big|_{\tau=0}=c_{A0}(z) \qquad z=[0,L] \tag{9-88}$$

Case 3 will be discussed separately in Section 9.5.3.

9.5.2 Separation of variables

The system of partial differential equations is solved using a separation of variables technique. By the principle of the technique, $\phi(z,\tau)$ is assumed to be separable into independent functions of position and transformed time, of the form

$$\phi(z,\tau)=\Psi(z)\cdot\Gamma(\tau) \tag{9-89}$$

Thus the dynamic equation (9-84) becomes

$$\frac{1}{\Psi(z)}\frac{d^2\Psi}{dz^2}=\frac{1}{\Gamma(\tau)}\frac{d\Gamma}{d\tau}=-\beta^2 \tag{9-90}$$

As the left-hand side of the equation is a function of z only while the right-hand is a function of τ only, this is only satisfied if they both equal a constant. Allow $-\beta^2$ to be the arbitrary constant. β is termed the separation variable.

9.5.2.1 Solution to the $\Gamma(\tau)$ problem

The $\Gamma(\tau)$ problem is expressed as

$$\frac{d\Gamma}{d\tau}=-\beta^2\Gamma(\tau) \tag{9-91}$$

which the following is a possible solution:

$$\Gamma(\tau)=\exp\left(-\beta^2\tau\right) \tag{9-92}$$

9.5.2.2 *Solution to the spatial problem*

The spatial problem is expressed as

$$\frac{d^2\Psi}{dz^2} = -\beta^2\Psi(z) \tag{9-93}$$

with the boundary conditions of

$$\Psi(z) = 0 \qquad z = L \tag{9-94}$$

$$\frac{d\Psi}{dz} = 0 \qquad z = 0 \tag{9-95}$$

Equation (9-93) has a solution of the general form of

$$\Psi(z) = A\cdot\sin(\beta z) + B\cdot\cos(\beta z) \tag{9-96}$$

From equation (9-95)we see that $A = 0$ and B may be arbitrarily chosen. Without loss of generality, set $B = 1$. Thus the solution to the spatial problem is

$$\Psi_n(z) = \cos(\beta_n z) \tag{9-97}$$

where $\beta_n : n = 1,2,3...$ are given by the positive roots defined by the boundary condition in equation (9-94), that is,

$$\cos(\beta_n L) = 0$$

$$\therefore \quad \beta_n = \frac{\pi}{L}\left(n - \tfrac{1}{2}\right) \tag{9-98}$$

9.5.2.3 *Initial conditions*

The complete solution for the system involves $\phi(z,\tau)$ being constructed by a linear superposition of the solutions

$$\begin{aligned}\phi(z,\tau) &= \sum_{n=1}^{\infty}\delta_n\cdot\Psi(\beta_n,z)\cdot\Gamma(\beta_n,\tau)\\ &= \sum_{n=1}^{\infty}\delta_n\cdot\cos(\beta_n z)\exp\left(-\beta_n{}^2\tau\right)\end{aligned} \tag{9-99}$$

where δ_n are constant coefficients used to satisfy the initial boundary conditions. The values of δ_n can be explicitly determined by solving equation (9-99) at $t = \tau = 0$

$$\phi(z,\tau)\big|_{t=0} = \sum_{n=1}^{\infty}\delta_n\cdot\cos(\beta_n z) = c_{A0}(z) \tag{9-100}$$

and using the property that the cosine functions are orthogonal

$$\int_0^L \cos(\beta_m z)\cdot\cos(\beta_n z)\,dz = \begin{cases} 0 & \text{for } m \neq n \\ \left(\dfrac{L}{2}\right) & \text{for } m = n \end{cases} \tag{9-101}$$

Operate both sides of equation (9-100) with the operator $\int_0^L \cos(\beta_r z)\,dz$ for an arbitrary r. Thus these we obtain

$$\sum_{n=1}^{\infty}\delta_n \int_0^L \cos(\beta_r z)\cdot\cos(\beta_n z)\,dz = \int_0^L \cos(\beta_r z)\cdot c_{A0}(z)\,dz$$

$$\therefore\quad \delta_n = \left(\frac{2}{L}\right)\int_0^L \cos(\beta_n z)c_{A0}(z)\,dz \tag{9-102}$$

Substituting in these coefficients, the generalized solution becomes

$$\phi(z,\tau) = \frac{2}{L}\sum_{n=1}^{\infty}\exp\left(-\beta_n^{\,2}\tau\right)\cos(\beta_n z)\int_0^L \cos(\beta_n z')c_{A0}(z')\,dz' \tag{9-103}$$

With constant c_{A0}, this simplifies to

$$\phi(z,\tau) = -\frac{2c_{A0}}{L}\sum_{n=1}^{\infty}\exp\left(-\beta_n^{\,2}\tau\right)\frac{(-1)^n}{\beta_n}\cos(\beta_n z) \tag{9-104}$$

where

$$\beta_n = \frac{\pi}{L}\left(n-\tfrac{1}{2}\right) \qquad n = 1,2,3\ldots \tag{9-105}$$

This solution is then inverted back into concentration terms using equation (9-8). The surface flux is determined by differentiating equation (9-8) with respect to the space variable and multiplying by the diffusion coefficient

$$\begin{aligned} j_A(t) &= -D_{A(eff)}\left.\frac{\partial c_A}{\partial z}\right|_{z=L} \\ &= -D_{A(eff)}\exp\left[-\int_0^t \frac{\rho_b \dfrac{dK_{sa}}{dt'}}{\varepsilon + \rho_b K_{sa}(t')}\,dt'\right]\left.\frac{\partial \phi}{\partial z}\right|_{z=L} \end{aligned} \tag{9-106}$$

where

$$\left.\frac{\partial\phi}{\partial z}\right|_{z=L} = \frac{2}{L}\sum_{n=1}^{\infty}\exp\left(-\beta_n{}^2\tau\right)\left(-1\right)^n\beta_n\int_0^L\cos\left(\beta_n z'\right)c_{A0}\left(z'\right)dz' \tag{9-107}$$

With constant c_{A0} this simplifies to

$$\left.\frac{\partial\phi}{\partial z}\right|_{z=L} = -\frac{2c_{A0}}{L}\sum_{n=1}^{\infty}\exp\left(-\beta_n{}^2\tau\right) \tag{9-108}$$

9.5.3 Time-dependent boundary condition

9.5.3.1 Duhamel's principle

Applying a normalized variation of the variable transformations given in Section 9.5.1, where

$$c_A(z,t) = u(z,t)\cdot c_{A0}\exp\left[-\int_0^t \frac{\rho_b\frac{dK_{sa}}{dt'}}{\varepsilon+\rho_b K_{sa}(t')}dt'\right] \tag{9-109}$$

and

$$\tau = \int_0^t \frac{D_{A(eff)}}{\varepsilon+\rho_b K_{sa}(t')}dt' \tag{9-110}$$

the transformed equations for case 3 can be stated as

$$\frac{\partial u}{\partial\tau} = \frac{\partial^2 u}{\partial z^2} \qquad z = [0,L] \tag{9-111}$$

$$\left.u(z,\tau)\right|_{z=L} = 0 \qquad \tau > 0 \tag{9-112}$$

$$\left.u(z,\tau)\right|_{z=0} = \exp\left[\int_0^t \frac{\rho_b\frac{dK_{sa}}{dt}}{\varepsilon+\rho_b K_{sa}(t)}dt\right] \qquad \tau > 0 \tag{9-113}$$

$$\left.u(z,\tau)\right|_{\tau=0} = 1 \qquad z = [0,L] \tag{9-114}$$

Thus, the system is transformed into a simple diffusion equation with a time-dependent boundary condition.

Time-dependent boundary conditions problems may be solved using Duhamel's principle. According to this method, the solution will have to form

$$u(z,t) = f(0)w_1(z,t) + \int_0^t \left\{ w_2(z,t-t')\frac{df}{dt'} \right\} dt' \tag{9-115}$$

where $f(t')$ is the function value of the time-dependent boundary, that is,

$$f(t') = \exp\left[\int_0^{t'} \frac{\rho_b \dfrac{dK_{sa}}{dt}}{\varepsilon + \rho_b K_{sa}(t)} dt \right] \tag{9-116}$$

and $w_1(z,t)$ is the solution to the system equation with the time-dependent boundary set to unity, and $w_2(z,t)$ is the solution to the system equation with the time-dependent boundary set to unity and the initial conditions set to zero.

Let the first component of equation (9-115) be denoted by $u_1(z,t)$ and the second, integral component of this equation be denoted by $u_2(z,t)$. The first component gives the contribution of the initial conditions and initial boundary condition to the system. The second component gives the contribution of the sum of pulse releases at the time-dependent boundary, over the integrated time period. These two contributions are added to provide the final solution.

9.5.3.2 Nonhomogeneous problem solution

Equation (9-115) requires the solution to the following two nonhomogenous systems. These are easily solved in terms of the transformed time variable, τ. The problems are stated as follows:

Nonhomogeneous problem 1:

$$\frac{\partial w_1}{\partial \tau} = \frac{\partial^2 w_1}{\partial z^2} \qquad z \in [0, L], \quad t > 0 \tag{9-117}$$

$$w_1(z,\tau) = 0 \qquad z = L, \quad t > 0 \tag{9-118}$$

$$w_1(z,\tau) = 1 \qquad z = 0, \quad t > 0 \tag{9-119}$$

$$w_1(z,\tau) = 1 \qquad z \in [0, L], \quad t = 0 \tag{9-120}$$

Nonhomogeneous problem 2:

$$\frac{\partial w_2}{\partial \tau} = \frac{\partial^2 w_2}{\partial z^2} \qquad z \in [0, L], \quad t > 0 \tag{9-121}$$

$$w_2(z,\tau) = 0 \qquad z = L, \quad t > 0 \tag{9-122}$$

$$w_2(z,\tau)=1 \qquad z=0,\quad t>0 \tag{9-123}$$

$$w_2(z,\tau)=0 \qquad z\in[0,L],\quad t=0 \tag{9-124}$$

It can be easily seen that each of these two problems has the following steady-state solution

$$w_{ss}(z)=1-\frac{z}{L} \qquad z\in[0,L] \tag{9-125}$$

By a separation of variables technique, as described in Section 9.5.2, the general solution for the two problems may be derived to be

$$w_i(z,\tau)=w_{ss}(z)+\frac{2}{L}\sum_{n=1}^{\infty}\sin\left(\frac{n\pi}{L}z\right)\int_0^L\left[w_i(z',0)-w_{ss}(z')\right]\sin\left(\frac{n\pi}{L}z'\right)dz' \tag{9-126}$$

Evaluating the above equation, we obtain the following explicit solutions:

$$\begin{aligned}w_1(z,\tau)&=\left(1-\frac{z}{L}\right)+\frac{2}{L}\sum_{n=1}^{\infty}\exp\left\{-\left(\frac{n\pi}{L}\right)^2\tau\right\}\sin\left(\frac{n\pi}{L}z\right)\int_0^L\left(\frac{z'}{L}\right)\sin\left(\frac{n\pi}{L}z'\right)dz'\\&=\left(1-\frac{z}{L}\right)-\frac{2}{\pi}\sum_{n=1}^{\infty}\frac{(-1)^n}{n}\sin\left(\frac{n\pi}{L}z\right)\exp\left\{-\left(\frac{n\pi}{L}\right)^2\tau\right\}\end{aligned} \tag{9-127}$$

$$\begin{aligned}w_2(z,\tau)&=\left(1-\frac{z}{L}\right)+\frac{2}{L}\sum_{n=1}^{\infty}\exp\left\{-\left(\frac{n\pi}{L}\right)^2\tau\right\}\sin\left(\frac{n\pi}{L}z\right)\int_0^L\left(\frac{z'}{L}-1\right)\sin\left(\frac{n\pi}{L}z'\right)dz'\\&=\left(1-\frac{z}{L}\right)-\frac{2}{\pi}\sum_{n=1}^{\infty}\frac{1}{n}\sin\left(\frac{n\pi}{L}z\right)\exp\left\{-\left(\frac{n\pi}{L}\right)^2\tau\right\}\end{aligned} \tag{9-128}$$

9.5.3.3 *General solution*

Inverting the time transformation on the $w_i(z,\tau)$ expressions derived above and using $f(0)=1$ we obtain the following expressions:

$$u_1(z,t)=\left(1-\frac{z}{L}\right)-\frac{2}{\pi}\sum_{n=1}^{\infty}\frac{(-1)^n}{n}\sin\left(\frac{n\pi}{L}z\right)\exp\left\{-\left(\frac{n\pi}{L}\right)^2\int_0^t\frac{D_{A(eff)}}{\varepsilon+\rho_bK_{sa}(t)}dt\right\} \tag{9-129}$$

$$u_2(z,t)=\int_{t'=0}^{t}\left\{\left[\left(1-\frac{z}{L}\right)-\frac{2}{\pi}\sum_{n=1}^{\infty}\frac{1}{n}\sin\left(\frac{n\pi}{L}z\right)\exp\left\{-\left(\frac{n\pi}{L}\right)^2\int_{t'}^{t}\frac{D_{A(eff)}}{\varepsilon+\rho_bK_{sa}(t)}dt\right\}\right]\frac{df}{dt'}\right\}dt' \tag{9-130}$$

where

$$\frac{df}{dt'} = \left(\frac{\rho_b \dfrac{dK_{sa}}{dt'}}{\varepsilon + \rho_b K_{sa}(t')} \right) \exp\left[\int_0^{t'} \frac{\rho_b \dfrac{dK_{sa}}{dt}}{\varepsilon + \rho_b K_{sa}(t)} dt \right] \tag{9-131}$$

and

$$u(z,t) = u_1(z,t) + u_2(z,t) \tag{9-132}$$

The concentration is then determined by equation (9-109).

For the partial differentials of $u_1(z,t)$ and $u_2(z,t)$, the two subexpressions which contain the space variable are differentiated at the surface $z = L$ to give the following:

$$\left.\frac{\partial}{\partial z}\left(1 - \frac{z}{L}\right)\right|_{z=L} = -\frac{1}{L} \tag{9-133}$$

$$\begin{aligned} \left.\frac{\partial}{\partial z}\sin\left(\frac{n\pi}{L}z\right)\right|_{z=L} &= \left(\frac{n\pi}{L}\right)\cos(n\pi) \\ &= \left(\frac{n\pi}{L}\right)(-1)^n \end{aligned} \tag{9-134}$$

Thus the partial differential of equations (9-129) and (9-130) with respect to z give

$$\left.\frac{\partial u_1}{\partial z}\right|_{z=L} = -\frac{1}{L} - \frac{2}{L}\sum_{n=1}^{\infty} \exp\left\{ -\left(\frac{n\pi}{L}\right)^2 \int_0^t \frac{D_{A(eff)}}{\varepsilon + \rho_b K_{sa}(t)} dt \right\} \tag{9-135}$$

$$\frac{\partial u_2}{\partial z} = \frac{1}{L} \int_{t'=0}^{t} \left\{ \left[-1 - 2\sum_{n=1}^{\infty} \exp\left\{ -\left(\frac{n\pi}{L}\right)^2 \int_{t'}^{t} \frac{D_{A(eff)}}{\varepsilon + \rho_b K_{sa}(t)} dt \right\} \right] \frac{df}{dt'} \right\} dt' \tag{9-136}$$

Hence, the flux at the surface is given by

$$j_A(t) = -D_{A(eff)} c_{A0} \exp\left[-\int_0^t \frac{\rho_b \dfrac{dK_{sa}}{dt}}{\varepsilon + \rho_b K_{sa}(t)} dt \right] \left(\left.\frac{\partial u_1}{\partial z}\right|_{z=L} + \left.\frac{\partial u_2}{\partial z}\right|_{z=L} \right) \tag{9-137}$$

References

Crank, J. (1975) *The Mathematics of Diffusion*, Clarendon Press, Oxford.

Farlow, S.J. (1993) *Partial Differential Equations for Scientists and Engineers*, Dover Publications, New York.

Valsaraj, K.T., Choy, B., Ravikrishna, R., Reible, D.D., Thibodeaux, L.J., Price, C.B., Brannon, J.M., Myers, T.E. (1997) Air emissions from exposed, contaminated sediments and dredged materials 1. Experimental data in laboratory microcosms and mathematical modelling, *J. Haz. Materials*, **54**, 65-87.

10 Constant flux liquid evaporation

10.1 Introduction

A final problem for consideration is also based upon vapor transport in soils. During the initial stages of liquid evaporation from a soil, capillarity will tend to maintain a nearly uniform liquid saturation. This results in a constant evaporative flux for a finite period until capillarity can no longer sustain a nearly uniform liquid saturation. Section 10.2 gives the mathematical development of a model for predicting the duration of the regime of constant flux.

10.2 Analysis and Development

Consider a semi-infinite sediment medium. Initially it has liquid phase uniformly distributed in it. Mass transport at the surface is limited by external conditions, i.e., constant flux exists at the boundary. It is assumed that the apparent effective diffusion of the liquid in the unsaturated sediment be approximately constant during this initial constant flux evaporation stage.

Assume the liquid flow in the porous medium may be represented by the standard diffusion equation

$$\frac{\partial \rho_A}{\partial t} = D_{L(app)} \frac{\partial^2 \rho_A}{\partial z^2} \tag{10-1}$$

where the apparent effective liquid diffusivity, $D_{L(eff)}$, is calculated from the relative permeability and capillary suction forces of the porous medium usually at the initial conditions.

$$D_{L(app)}(\theta) = -\left(\frac{K^{sat}}{\mu_{liq}}\right) k_{r,liq}(\theta) \frac{dp_c}{d\theta} \tag{10-2}$$

The diffusive flux is given by definition

$$j_A(z,t) = -D_{L(app)} \frac{\partial \rho_A}{\partial z} \tag{10-3}$$

Differentiate equation (10-1) with respect to the space variable, and rearranging we obtain

$$\frac{\partial j_A}{\partial t} = D_{L(app)} \frac{\partial^2 j_A}{\partial z^2} \tag{10-4}$$

The above equation is known as the *flux formulation* of the diffusion equation. Note that it has the same form as the original diffusion equation. This form is useful in solving problems with flux-type boundary conditions.

The system formulation is given by

$$\frac{\partial j_A}{\partial t} = D_{L(app)} \frac{\partial^2 j_A}{\partial z^2} \quad \text{for } z \in [0,\infty), t > 0 \quad (10\text{-}5)$$

$$j_A(z,t) = -k_a \cdot c^{sat} = \text{constant} \quad \text{for } z = 0, t > 0 \quad (10\text{-}6)$$

$$j_A(z,t) = 0 \quad \text{for } t = 0, z \in [0,\infty) \quad (10\text{-}7)$$

This system is illustrated by the following diagram:

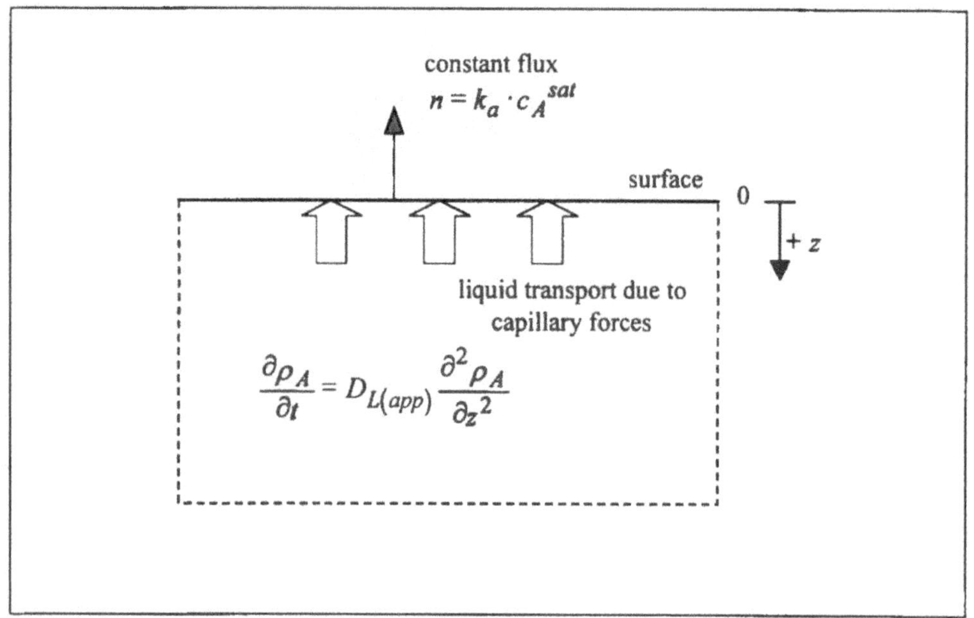

Figure 10-1 System with volatile liquid evaporation at a constant flux rate.

Define a new variable by the transformation

$$N(z,t) = j_A(z,t) + k_a \cdot c^{sat} \quad (10\text{-}8)$$

Then the problem can be expressed in the form

$$\frac{\partial N}{\partial t} = D_{L(app)} \frac{\partial^2 N}{\partial z^2} \quad \text{for } z \in [0,\infty), t > 0 \quad (10\text{-}9)$$

$$N(z,t) = 0 \quad \text{for } z = 0, t > 0 \quad (10\text{-}10)$$

$$N(z,t) = k_a \cdot c^{sat} \quad \text{for } t = 0, z \in [0,\infty) \quad (10\text{-}11)$$

This problem has a semi-infinite medium solution of the form given in Section 3.2.1.

$$N(z,t) = \left(k_a \cdot c^{sat}\right) \text{erf}\left\{\frac{z}{\sqrt{4D_{L(app)}t}}\right\} \quad (10\text{-}12)$$

Transforming this back into the flux variable we obtain

$$
\begin{aligned}
j_A(z,t) &= -\left(k_a \cdot c^{sat}\right)\left[1-\operatorname{erf}\left\{\frac{z}{\sqrt{4D_{L(app)}t}}\right\}\right] \\
&= -\left(k_a \cdot c^{sat}\right)\operatorname{erfc}\left\{\frac{z}{\sqrt{4D_{L(app)}t}}\right\}
\end{aligned}
\tag{10-13}
$$

Once the flux is known, the concentration profile is obtained by integrating this by the space variable

$$
\rho_A(z,t) = -\left(\frac{k_a \cdot c^{sat}}{D_{L(app)}}\right)\int_z^{\infty} \operatorname{erfc}\left\{\frac{z'}{\sqrt{4D_{L(app)}t}}\right\}dz' \tag{10-14}
$$

This gives the result of

$$
\rho_A(z,t) = \rho_0 - \frac{2 \cdot k_a \cdot c^{sat}}{D_{L(app)}}\left[\sqrt{D_{L(app)}\frac{t}{\pi}}\exp\left(\frac{-z^2}{4D_{L(app)}t}\right) - \frac{z}{2}\operatorname{erfc}\left(\frac{z}{\sqrt{4D_{L(app)}t}}\right)\right] \tag{10-15}
$$

Thus the concentration at the surface is given by

$$
\rho_A(0,t) = \rho_0 - \left[\frac{2 \cdot k_a \cdot c^{sat}}{\sqrt{\pi \cdot D_{L(app)}}}\right]\sqrt{t} \tag{10-16}
$$

The period of constant flux may be assumed to end when the surface concentration becomes zero. Thus the time during which constant flux from the surface occurs until

$$
t_{end} = \frac{\rho_0^{\,2} \cdot \pi \cdot D_{L(app)}}{4 \cdot k_a^{\,2} \cdot \left(c^{sat}\right)^2} \tag{10-17}
$$

References

Carslaw, H.S., Jaeger, J.C. (1959) *Conduction of Heat in Solids*, Clarendon Press, Oxford.

Thibodeaux, L.J. (1996) *Environmental Chemodynamics*, 2nd ed., John Wiley & Sons, New York.

Appendix

A. Error function

The error function of argument x is defined as

$$\mathrm{erf}(x) = \frac{2}{\sqrt{\pi}} \int_0^x e^{-\eta^2} d\eta \tag{1}$$

Therefore we have

$$\mathrm{erf}(\infty) = 1 \tag{2}$$

$$\mathrm{erf}(-x) = -\mathrm{erf}(x) \tag{3}$$

Table A-1: Numerical values of the error function

X	erf(x)	x	erf(x)	x	erf(x)
0.00	0.00000	0.25	0.27633	1.30	0.93401
0.01	0.01128	0.30	0.32863	1.35	0.94376
0.02	0.02256	0.35	0.37938	1.40	0.95229
0.03	0.03384	0.40	0.42839	1.45	0.95970
0.04	0.04511	0.45	0.47548	1.50	0.96611
0.05	0.05637	0.50	0.52050	1.55	0.97162
0.06	0.06762	0.55	0.56332	1.60	0.97635
0.07	0.07886	0.60	0.60386	1.65	0.98038
0.08	0.09008	0.65	0.64203	1.70	0.98379
0.09	0.10128	0.70	0.67780	1.75	0.98667
0.10	0.11246	0.75	0.71116	1.80	0.98909
0.11	0.12362	0.80	0.74210	1.85	0.99111
0.12	0.13476	0.85	0.77067	1.90	0.99279
0.13	0.14587	0.90	0.79691	1.95	0.99418
0.14	0.15695	0.95	0.82089	2.00	0.99532
0.15	0.16800	1.00	0.84270	2.20	0.99814
0.16	0.17901	1.05	0.86244	2.40	0.99931
0.17	0.18999	1.10	0.88021	2.60	0.99976
0.18	0.20094	1.15	0.89612	2.80	0.99992
0.19	0.21184	1.20	0.91031	3.00	0.99998
0.20	0.22270	1.25	0.92290		

The complimentary error function of argument x is defined as

$$\operatorname{erfc}(x) = 1 - \operatorname{erf}(x) = \frac{2}{\sqrt{\pi}} \int_x^{\infty} e^{-\eta^2} d\eta \tag{4}$$

The derivative of the error function is given as

$$\frac{d}{dx} \operatorname{erf}(x) = \frac{2}{\sqrt{\pi}} e^{-x^2} \tag{5}$$

Evaluation of the error function may be done using the following approximate curve fit by Chebyshev (given below in Fortran code):

```
      PROGRAM ERF_CALC
      REAL A,ERFA
      WRITE (*,10)
      READ (*,*) A
      ERFA=ERF(A)
      WRITE (*,20) A,ERFA
10    FORMAT(' ENTER X')
20    FORMAT(' ERF(',F10.5,')= ',F10.5)
      STOP
      END

      FUNCTION ERF(X)
      REAL ERF,X
      REAL Z,T
      Z=ABS(X)
      T=1./(1.+0.5*Z)
         ERF=1-(T*EXP(-Z*Z-1.26551223+T*(1.00002368+T*(.37409196+
     & T*(.09678418+T*(-.18628806+T*(.27886807+T*(-1.13520398+
     & T*(1.48851587+T*(-.82215223+T*.17087277)))))))))))
      IF (X.LT.0.) ERF=-ERF
      RETURN
      END
```

B. Laplace transformation

The Laplace transform is an important integral transform with a number of useful properties. The transform maps a point in the time domain, variable t, linearly to a complex mathematical domain called the Laplace domain, variable s. It is defined in the following manner:

$$\mathscr{L}[f] = F(s) = \int_0^{\infty} f(t) \cdot e^{-st} dt \tag{1}$$

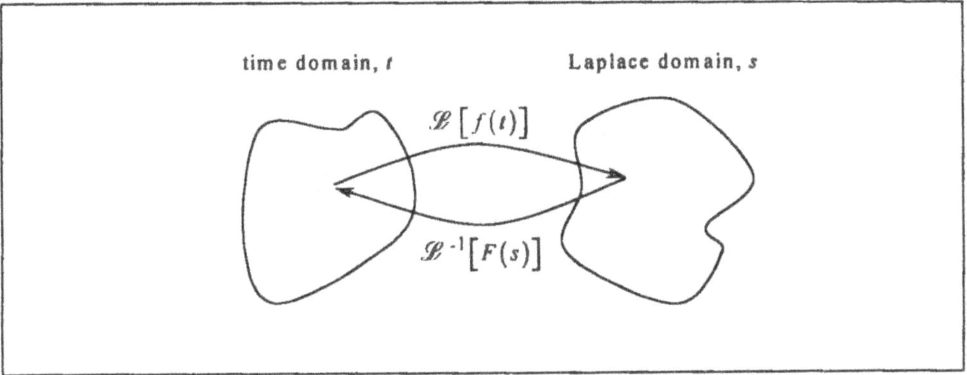

Figure B-1 Transformation and inverse transformation from the time and Laplace domains.

Properties of the Laplace transformation

The property of the Laplace transform of most interest to solving advection-diffusion equations is the transforms of the partial derivatives

$$\mathscr{L}\left[\frac{\partial c_A}{\partial t}\right] = \int_0^{\infty} \frac{\partial c_A}{\partial t} \cdot e^{-st} dt = s \cdot C(z,s) - c_A(z,0) \tag{2}$$

$$\mathscr{L}\left[\frac{\partial c_A}{\partial z}\right] = \int_0^{\infty} \frac{\partial c_A}{\partial z} \cdot e^{-st} dt = \frac{dC}{dz}(z,s) \tag{3}$$

$$\mathscr{L}\left[\frac{\partial^2 c_A}{\partial z^2}\right] = \int_0^{\infty} \frac{\partial^2 c_A}{\partial z^2} \cdot e^{-st} dt = \frac{d^2 C}{dz^2}(z,s) \tag{4}$$

where $\mathscr{L}[c_A(z,t)] = C(z,s)$.

With the above properties, it can be seen that the partial differential equations in time and space can now be expressed as a ordinary differential equation in space only.

The other property of importance to solving advection-diffusion systems is the linearity of the transform

$$\begin{aligned}\mathscr{L}[A\cdot f(t)+B\cdot g(t)] &= \int_0^\infty [A\cdot f(t)+B\cdot g(t)]\cdot e^{-st}dt \\ &= A\int_0^\infty f(t)\cdot e^{-st}dt + B\int_0^\infty g(t)\cdot e^{-st}dt \\ &= A\cdot\mathscr{L}[f(t)]+B\cdot\mathscr{L}[g(t)]\end{aligned} \tag{5}$$

This property assists with the inversion of the Laplace transforms back to the time domain.

Inversion of the Laplace transformation

The inverse of the Laplace transform is defined by the following contour integral in the complex plane:

$$\mathscr{L}^{-1}[F(s)] = f(t) = \frac{1}{2\pi i}\int_{c-i\infty}^{c+i\infty} F(s)\cdot e^{st}ds \tag{6}$$

where c is a very large number which encompasses all the features of $F(s)$.

Alternatively, the following table of transforms and their inverses may be applied.

Table B-1 Table of Laplace transforms

	$f(t)=\mathscr{L}^{-1}[F(s)]$	$F(s)=\mathscr{L}[f(t)]$
1.	$t^n \quad n=1,2,3\ldots$	$\frac{n!}{s^{n+1}} \quad s>0$
2.	1	$\frac{1}{s} \quad s>0$
3.	$\exp(at)$	$\frac{1}{s-a} \quad s>a$
4.	$\sin(at)$	$\frac{a}{s^2+a^2} \quad s>0$
5.	$\cos(at)$	$\frac{s}{s^2+a^2} \quad s>0$

6.	$\sinh(at)$	$\frac{a}{s^2-a^2}\quad s>\lvert a\rvert$

	$f(t)=\mathscr{L}^{-1}[F(s)]$	$F(s)=\mathscr{L}[f(t)]$
7.	$\cosh(at)$	$\frac{s}{s^2-a^2}\quad s>\lvert a\rvert$
8.	$H(t-t_0)$	$\frac{1}{s}\exp(-s\cdot t_0)\quad s>0$
9.	$\operatorname{erf}\left(\frac{t}{2a}\right)$	$\frac{1}{s}\exp\left(a^2s^2\right)\operatorname{erfc}(as)$
10.	$\operatorname{erfc}\left(\frac{a}{\sqrt{4t}}\right)$	$\frac{1}{s}\exp\left(-a\sqrt{s}\right)$
11.	$\delta(t-t_0)$	$\exp(-s\cdot t_0)$
12.	$\frac{1}{\sqrt{\pi t}}\exp\left(\frac{-a^2}{4t}\right)$	$\frac{1}{\sqrt{s}}\exp\left(\frac{-a}{s}\right)\quad a\ge 0$
13.	$-\exp(ak)\exp\left(a^2t\right)\operatorname{erfc}\left[\left(\frac{k}{2\sqrt{t}}\right)+a\sqrt{t}\right]$ $\ldots+\operatorname{erf}\left(\frac{k}{2\sqrt{t}}\right)\qquad k\ge 0$	$\frac{a\exp\left(-k\sqrt{s}\right)}{s\left(a+\sqrt{s}\right)}$
14.	$\exp(ak)\exp\left(a^2t\right)\operatorname{erfc}\left[\left(\frac{k}{2\sqrt{t}}\right)+a\sqrt{t}\right]$ $k\ge 0$	$\frac{\exp\left(-k\sqrt{s}\right)}{\sqrt{s}\left(a-\sqrt{s}\right)}$
15.	$\frac{1}{2}\exp\left(a^2t\right)\left\{\exp(-ka)\operatorname{erfc}\left[\frac{k}{2t}-a\sqrt{t}\right]\right.$ $\left.+\exp(ka)\operatorname{erfc}\left[\frac{k}{2t}+a\sqrt{t}\right]\right\}$	$\frac{\exp\left(-k\sqrt{s}\right)}{s+a^2}$
16.	$-a\frac{1}{\sqrt{\pi t}}\exp\left(a^2t\right)\operatorname{erfc}\left(a\sqrt{t}\right)$	$\frac{1}{\sqrt{s}+a}$

C. Roots of transcendental equations

Eigenvalues are required for the numerical evaluation of several models. These eigenvalues are determined by roots of transcendental functions of various forms. All the roots within a specified range of significance need to be found. Although most of the numerical analyses of this publication's models were performed in specialized mathematical software, the following FORTRAN subroutines may be used in the numerical evaluation of these roots.

Common variables

```
B(n)   array of eigenvalues
BNUM   number of eigenvalues
BMAX   upper bound of the range of significance
C(n)   array of coefficients to the transcendental function
XTOL   numerical tolerance for the bisection algorithm
BOUNDS(n)    array of search bounds
BOUNDSNUM    number of search bounds
```

Transcendental function $\beta_n \tan(C_1\beta_n) - C_2 = 0$

```
        SUBROUTINE CALC_EIGENVALUES(C,BMAX,B)
        PARAMETER (XTOL=1.0E-4)
        PARAMETER (PI=3.14159265358979)
        REAL C(2),BMAX,B(1000)
        INTEGER BNUM
        REAL BOUNDS(1000),X,UBD,LBD,MIDPT
        INTEGER N,BOUNDSNUM
C INITIALISE VARIABLES
        BNUM=0
        BOUNDSNUM=1
        BOUNDS(BOUNDSNUM)=0.
        N=1
C CALCULATE BOUNDS
        DO WHILE (BOUNDS(BOUNDSNUM).LT.BMAX)
            BOUNDSNUM=BOUNDSNUM+1
            X=BOUNDSNUM
            BOUNDS(BOUNDSNUM)=PI*(X-0.5)/c(1)
        ENDDO
C CALCULATE EIGENVALUES
        DO WHILE (N.LT.BOUNDNUM)
            BNUM=BNUM+1
            UBD=BOUNDS(N+1)
            LBD=BOUNDS(N)
            DO WHILE ((UBD-LBD).GT.XTOL)
                MIDPT=0.5*(UBD-LBD)+LBD
                B(BNUM)=MIDPT*TAN(C(1)*MIDPT)-C(2)
                IF (B(BNUM).EQ.0.0)
                    UBD=MIDPT
                    LBD=MIDPT
                ELSEIF(B(NUM)*UBD.GT.0.0)
                    UBD=MIDPT
                ELSE
                    LBD=MIDPT
```

```
            ENDIF
            B(BNUM)=UBD
        ENDDO
        N=N+1
    ENDDO
    RETURN
    END
```

Transcendental function $C_1\cos(C_2\beta_n)+C_3\cos(C_4\beta_n)=0$

```
      SUBROUTINE CALC_EIGENVALUES(C,B,BOUNDS,BOUNDSNUM)
C BOUNDS ARRAY HAS BEEN CALCULATED AND SORTED
C IN A PREVIOUS CALLED SUBROUTINE
      PARAMETER (XTOL=1.0E-4)
      REAL C(4),BMAX,B(1000),BOUNDS(1000)
      INTEGER BNUM,BOUNDSNUM
      REAL X1,X2,UBD,LBD,MIDPT
      INTEGER N
C INITIALISE VARIABLES
      BNUM=0
      N=1
C CALCULATE EIGENVALUES
      DO WHILE (N.LT.BOUNDNUM)
          UBD=BOUNDS(N+1)
          LBD=BOUNDS(N)
          X1=C(1)*COS(C(2)*UBD)+C(3)*COS(C(4)*UBD)
          X2=C(1)*COS(C(2)*LBD)+C(3)*COS(C(4)*LBD)
          IF (X1.EQ.0.0)
              BNUM=BNUM+1
              B(BNUM)=LBD
          ELSEIF (X1*X2.LT.0.0)
              BNUM=BNUM+1
              DO WHILE (ABS(UBD-LBD).GT.XTOL)
                  MIDPT=0.5*(UBD-LBD)+LBD
                  B(BNUM)=C(1)*COS(C(2)*MIDPT)
                  B(BNUM)=BNUM+C(3)*COS(C(4)*MIDPT)
                  IF (B(BNUM).EQ.0.0)
                      UBD=MIDPT
                      LBD=MIDPT
                  ELSEIF(B(NUM)*UBD.GT.0.0)
                      UBD=MIDPT
                  ELSE
                      LBD=MIDPT
                  ENDIF
              ENDDO
              B(BNUM)=UBD
          ENDIF
          N=N+1
      ENDDO
      RETURN
      END
```

D. Predicting the diffusion coefficient in vapors

Diffusion coefficients in vapor can be estimated by the following equation by Fuller, Schettler, and Giddings (1966).

$$D^V = \frac{10^{-7} T^{1.75} \sqrt{\frac{1}{M_a} + \frac{1}{M_b}}}{P\left[\left(\sum_a v_i\right)^{\frac{1}{3}} + \left(\sum_b v_i\right)^{\frac{1}{3}}\right]^2}$$

where D^V = vapor diffusivity [m²/s]

T = temperature [K]

M_a, M_b = molecular weights of components *a* and *b* [g/mol]

P = total pressure [atm]

$\sum_a v_i, \sum_b v_i$ = sum of *diffusion volume coefficients* for *a* and *b*, as given below

Table D-1 Diffusion volume coefficients (Fuller *et al.*, 1966)

Atomic and structural diffusion volumes [cm³/mol]			
C	16.5	Cl	19.5*
H	1.98	S	17.0*
O	5.48	Aromatic or heterocyclic rings	-20.2
N	5.69*		

Diffusion volumes of simple molecules [cm³/mol]			
H_2	7.07	CO	18.9
D_2	6.70	CO_2	26.9
He	2.88	N_2O	35.9
N_2	17.9	NH_3	14.9
O_2	16.6	H_2	12.7
Air	20.1	CCl_2F_2	114.8*
Ne	5.59	SF_6	69.7*
Ar	16.1	Cl_2	37.7*
Kr	22.8	Br_2	67.2*
Xe	37.9*	SO_2	41.1*

* values are based on few data points

This method gives good results for nonpolar gases at low and moderate temperatures. The poorest correlation occurs with polar acids and glycols (Lyman et al., 1993).

References

Fuller, E.N., Schettler, P.D., Giddings, J.C. (1966) A new method of prediction of binary gas-phase diffusion coefficients, *Ind. Eng. Chem.*, **58**, 19-27.

Lyman, W.J., Reehl, W.F., Rosenblatt, D.H. (1993) *Handbook of Chemical Property Estimation: Environmental Behaviour of Organic Compounds*, American Chemical Society, New York.

E. Predicting the diffusion coefficient in liquids

Diffusion coefficients in liquids can be estimated by the following equation by Wilke and Chang (1955).

$$D^L = \frac{1.173\times10^{-16} T\sqrt{\phi\cdot M}}{\mu\cdot V_m{}^{0.6}}$$

where
D^L = liquid diffusivity [m²/s]
T = temperature [K]
ϕ = association factor for the solvent
M = molecular weight of the solvent [g/mol]
μ = viscosity of the solvent [Pa.s]
V_m = molar volume of the solute at its boiling point [m³/kmol]

The following parameter values are for water as the solvent.

ϕ = 2.26 (revised correlated value. 2.6 in original paper by Wilke and Chang, 1955)
M = 18.1 g/mol

Table E-1 Viscosity of water

Temp [°C]	μ [Pa.s]	Temp [°C]	μ [Pa.s]	Temp [°C]	μ [Pa.s]
0	1.787×10^{-3}				
1	1.728×10^{-3}	11	1.271×10^{-3}	21	0.978×10^{-3}
2	1.671×10^{-3}	12	1.235×10^{-3}	22	0.955×10^{-3}
3	1.618×10^{-3}	13	1.202×10^{-3}	23	0.933×10^{-3}
4	1.567×10^{-3}	14	1.169×10^{-3}	24	0.911×10^{-3}
5	1.519×10^{-3}	15	1.139×10^{-3}	25	0.890×10^{-3}
6	1.472×10^{-3}	16	1.109×10^{-3}	26	0.871×10^{-3}
7	1.428×10^{-3}	17	1.081×10^{-3}	27	0.851×10^{-3}
8	1.386×10^{-3}	18	1.053×10^{-3}	28	0.833×10^{-3}
9	1.346×10^{-3}	19	1.027×10^{-3}	29	0.815×10^{-3}
10	1.307×10^{-3}	20	1.002×10^{-3}	30	0.798×10^{-3}

The molar volume of the solute at its boiling point can be predicted by the group contribution method of Gambil (1958).

Table E-2 Structural contributions to molar volume at boiling point (Gambil, 1958)

Molecular volumes [m^3/kmol]					
Air	0.0299	H_2	0.0143	NO	0.0236
Br_2	0.0532	H_2O	0.0189	N_2O	0.0364
Cl_2	0.0484	H_2S	0.0329	O_2	0.0256
CO	0.0307	I_2	0.0715	SO_2	0.0448
CO_2	0.0340	N_2	0.0312		
COS	0.0515	NH_3	0.0258		

Atomic volumes [m^3/kmol]					
As	0.0305	H	0.0037	Si	0.0320
Bi	0.0480	Hg	0.0190	Sn	0.0423
Br	0.0270	I	0.0370	Ti	0.0357
C	0.0148	P	0.0270	V	0.0320
Cr	0.0274	Pb	0.0480	Zn	0.0204
F	0.0087	S	0.0256		
Ge	0.0345	Sb	0.0342		

Special contribution volumes [m^3/kmol]			
Cl		Three-membered ring	-0.0060
- terminal, as in R-Cl	0.0216	Four-membered ring	-0.0085
- medial as in R-CHCl-R	0.0246	Five-membered ring	-0.0115
N		Six-membered ring, as in benzene, cyclohexane, pridine	-0.0150
- double bonded	0.0156		
- triple bonded, as in nitriles	0.0162	Naphthalene ring	-0.0300
- in primary amines, RNH_2	0.0105	Anthracene ring	-0.0475
- in secondary amines, R_2NH	0.0120		
- in tertiary amines, R_3N	0.0108		
O, except noted below	0.0074		
- in methyl esters	0.0091		
- in methyl ethers	0.0099		
- in higher esters/ethers	0.0110		
- in acids	0.0120		
- in union with S, P, N	0.0083		

References

Wilke, C.R., Chang, P. (1955) Correlation of diffusion coefficients in dilute solutions, *AIChE J.*, **1**, 264.

Gambil, W.R. (1958) Predict diffusion coefficient, *Chem. Eng.*, **65**, 125.

F. Sample calculations of models using Mathcad™

Following are several Mathsoft's Mathcad worksheet templates and sample calculations for selected models described in this publication. These models include

1. Model 3-2: Concentration Profile and Surface Flux, Semi-infinite Layer Diffusion

2. Model 5-2: Concentration Profile and Surface Flux, Two-Layer Diffusion

3. Model 6-2: Concentration Profile and Surface Flux, Three-Layer Diffusion

A number of features of this mathematical analysis software package made it a useful tool for evaluation of the models. Some of the features of Mathcad version 7 that were taken advantage of included:

- Logical/intuitive worksheet layout (reads like a hand-written mathematical analysis rather than an archaic programming language)
- Dynamic variable declaration/memory allocation
- "Live" symbolic evaluation
- "Live" graphing of functions and results
- Limited programming capabilities for complex calculation algorithms
- Ability to create templates of worksheets

The template files and other sample calculations can be downloaded from the US EPA Hazardous Substance Research Center Web site:

http://www.hsrc.org

The files are resident in the area devoted to the South/Southwest regional center

http://www.hsrc.org/hsrc/html/south.html

Hazardous Substance Research Center
Louisiana State University

PROJECT DESCRIPTION Semi-infinite layer diffusion model (Model 3-2: concentration profile and surface flux)

Calculation of a surface flux profile from diffusion in a semi-infinite region
Ref: Contaminant Transport in Soils and Sediments, §3

CALCULATION BY Bruce Choy DATE 14 October 1997

$$-D_{A(eff)}\frac{\partial c_A}{\partial z}+k_a\cdot c_A=0$$

surface 0 +z

$$\frac{\partial c_A}{\partial t}=\left(\frac{D_{A(eff)}}{R_f}\right)\frac{\partial^2 c_A}{\partial z^2}$$

constant initial conditions

System parameters:

Diffusion coefficient, retardation factor and surface mass transfer coefficient, initial constant concentration as specified in the above diagram

$D:=5\cdot10^{-7}$ m²/s $R:=1000$ $k_a:=0.0005$ m/s $c_{A0}:=5\cdot10^{-4}$ kg/m³

Simulation parameter:

$m:=1..200$
$time_m:=m\cdot3600$ seconds

= array of time steps in which surface flux profile is calculated from

Calculate the surface flux profile

$\mathrm{erfc}(x) := 1 - \mathrm{erf}(x)$ = define the complimentary error function

note:

* Mathcad numerical limit to the error function, erf(x), occurs at high values of x; this is overcome by changing the model equation to the one with zero concentration surface boundary model at high values of time.
* The stabiltiy criterion is a function of surface mass transfer coefficient, thus the model degenerates to the zero surface concentration model for high values of k_a .

$$\mathrm{SurfaceFlux}(t) := \begin{cases} c_{A0}\cdot k_a\cdot \exp\left(\dfrac{k_a{}^2\cdot t}{D\cdot R}\right)\cdot\left(1-\mathrm{erf}\left(k_a\cdot\sqrt{\dfrac{t}{D\cdot R}}\right)\right) & \text{if } \left(\dfrac{k_a{}^2\cdot t}{D\cdot R}\right)\le 25 \\ c_{A0}\cdot\sqrt{\dfrac{D\cdot R}{\pi\cdot t}} & \text{otherwise} \end{cases}$$

$$\mathrm{Flux}_m := \mathrm{SurfaceFlux}\left(\mathrm{time}_m\right)$$

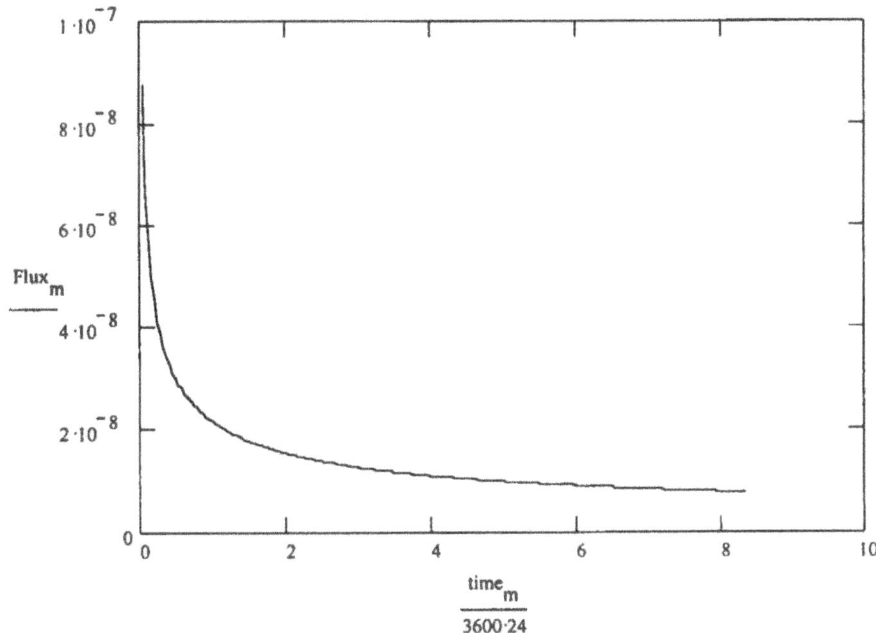

Calculate the final concentration profile

$L := 0.10$ m = total depth to calculate and graph

$z\_max\_div := 100$ $j := 0 .. z\_max\_div$ $\zeta_j := L \cdot \left(\frac{j}{z\_max\_div} \right)$

$$\mathrm{ConcCalc}(z,t) := \begin{cases} \text{if } \left(\frac{k_a^2 \cdot t}{D \cdot R} \right) \le 25 \\ \quad \alpha \leftarrow \frac{R \cdot z}{\sqrt{4 \cdot D \cdot R \cdot t}} \\ \quad c_{A0} \cdot \left(\mathrm{erf}(\alpha) + \exp\left(\frac{k_a \cdot z}{D} + \frac{k_a^2 \cdot t}{D \cdot R} \right) \cdot \mathrm{erfc}\left(\alpha + k_a \cdot \sqrt{\frac{t}{D \cdot R}} \right) \right) \\ c_{A0} \cdot \mathrm{erf}\left[\frac{z}{\sqrt{4 \cdot \left(\frac{D}{R} \right) \cdot t}} \right] \text{ otherwise} \end{cases}$$

$time_{max} := \max(time)$

$conc_j := \mathrm{ConcCalc}(\zeta_j, time_{max})$

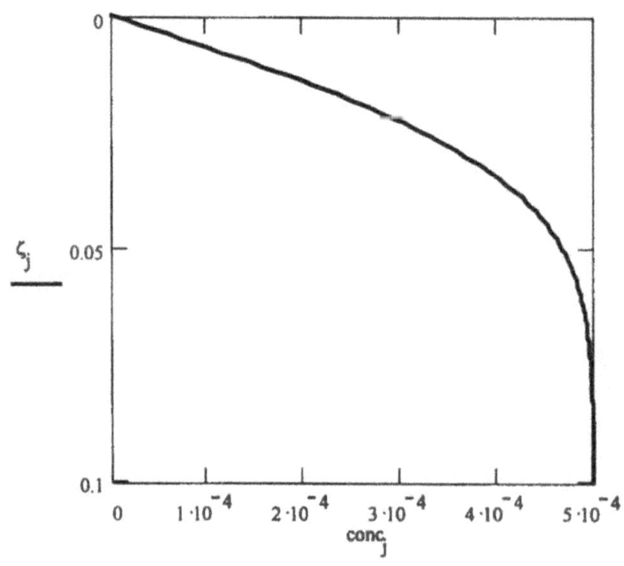

Hazardous Substance Research Center
Louisiana State University

PROJECT Two-layer diffusion model

DESCRIPTION (Model 5-3: surface flux profile and final concentration profile)

Calculation of a surface flux profile from diffusion in a two-layer composite
Ref: Contaminant Transport in Soils and Sediments, §5

CALCULATION BY Bruce Choy **DATE** 14 October 1997

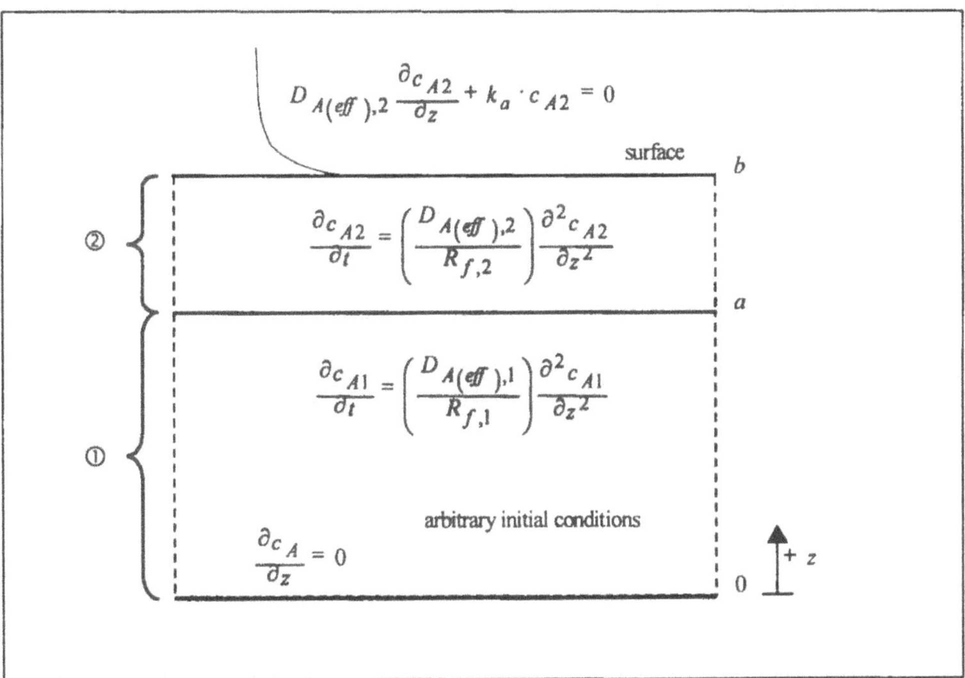

System parameters:

Diffusion coefficients, retardation factors, and layer locations as specified in th above diagram

$k_a := 0.001$ m/s = surface mass transfer coefficie

$D_2 := 5 \cdot 10^{-7}$ m²/s $R_2 := 600$ $b := 0.100$ m = surface layer dat

$D_1 := 1 \cdot 10^{-7}$ $R_1 := 1000$ $a := 0.070$ = bottom layer data

Simulation parameter:

$m := 1..200$ = array of time steps in which surface flux profile is calculated from

$time_m := (10 \cdot m + 10) \cdot 3600$ seconds

Initial concentration specification:

$$c0 := \begin{bmatrix} b & 0 \\ a & 10 \\ 0 & 0 \end{bmatrix}$$

notes:

1. The 'c0' array should be in order of descending depth where the left entry is 'z' and the right is 'conc' of the layer directly below it.
2. Specify the interface position, 'a', explicitly in the array (as this simplifies the integration process).
3. Top entry must contain the value of 'b' in the 'z' position and the concentration of the surface zone in the 'conc' position.
4. Bottom row must contain a zero-zero ordered pair.
5. This array is used to calculate the initialization integral.
6. Below is a plot which displays the initial concentration that has been specified by the 'c0' array.

```
plot_c0 := | x ← [c0_{0,0}  0]
           | for i ∈ 0..(rows(c0) − 2)
           |   | y_{0,0} ← c0_{i,0}
           |   | y_{0,1} ← c0_{i,1}
           |   | y_{1,0} ← c0_{i+1,0}
           |   | y_{1,1} ← c0_{i,1}
           |   | x ← stack(x, y)
           | x
```

$p := 0..(rows(plot\_c0) - 1)$

$InitConc_p := plot\_c0_{p,1}$

$depth_p := plot\_c0_{p,0}$

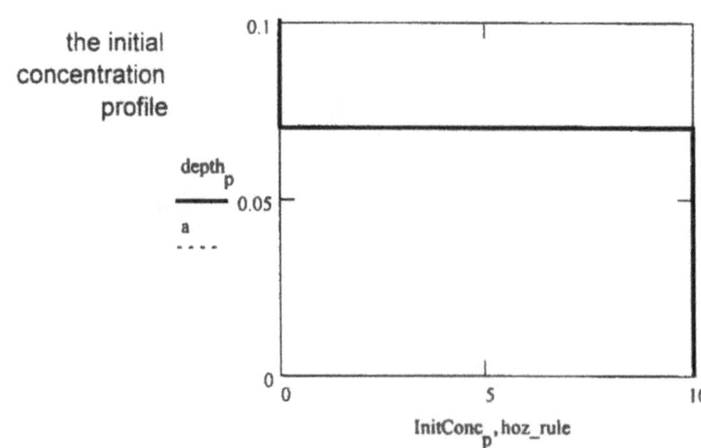

Calculate System Eigenvalues

System eigenvalues, β_n, are the roots of the following transcendental function, f(β)

$$i := 1..2 \qquad \alpha_i := \frac{D_i}{R_i} \qquad \gamma_i := \sqrt{D_i \cdot R_i}$$

$$C_1 := \gamma_1 + \gamma_2 \qquad C_2 := \frac{a}{\sqrt{\alpha_1}} + \frac{b-a}{\sqrt{\alpha_2}} \qquad C_3 := \gamma_1 - \gamma_2 \qquad C_4 := \frac{a}{\sqrt{\alpha_1}} - \frac{b-a}{\sqrt{\alpha_2}}$$

$$C_5 := k_a \cdot \left(\frac{\gamma_1}{\gamma_2} - 1\right) \qquad C_6 := \frac{a}{\sqrt{\alpha_1}} - \frac{b-a}{\sqrt{\alpha_2}} \qquad C_7 := -k_a \cdot \left(\frac{\gamma_1}{\gamma_2} + 1\right) \qquad C_8 := \frac{a}{\sqrt{\alpha_1}} + \frac{b-a}{\sqrt{\alpha_2}}$$

$$f(\beta) := C_1 \cdot \beta \cdot \sin(C_2 \cdot \beta) + C_3 \cdot \beta \cdot \sin(C_4 \cdot \beta) + C_5 \cdot \cos(C_6 \cdot \beta) + C_7 \cdot \cos(C_8 \cdot \beta)$$

Defining the range of significant eigenvalues, β=[0, bmax]

$\omega := 20$ = adjustable parameter to define range of significance

$$\beta_{max} := \frac{\omega \cdot \pi}{\min([\,|C_2|\ |C_4|\ |C_6|\ |C_8|\,])}$$ = range of significance β=[0, β_max)

$\beta_{max} = 0.011$

This subroutine calculates the roots of f(β) within the range $\beta=[0,\beta_{max}]$

notes:

1. The adjustable parameter ω is used to determine the range of significance $\beta=[0,\beta_{max}]$ (i.e., the range in which the eigenvalues are calculated).
2. Choose an 'ω' such that the final concentration profile has a high enough resolution frequency to yield an appropriate (smooth) curve. ω=20 is a good initial estimate.
3. Usually 10-20 eigenvalues, i.e., 'β's is a reasonable number for most 'time' variables.
4. Higher number of eigenvalues are required for short values of the 'time' variable (i.e., higher values ofω).
5. The values of the eigenvaules are calculated by the rigorous bracketing-bisection algorithm.
6. 'β-array' can be seen in the graph on the following page.

```
β_array := | bnds_0 ← 0
           | for i ∈ 1..2
           |   | nb_sub ← floor(β_max·C_2·i / π)
           |   | for j ∈ 0..nb_sub − 1
           |   |   bnd_sub_j ← (j + 0.5)·π / C_2·i
           |   | bnds ← stack(bnds, bnd_sub)
           | for i ∈ 3..4
           |   | nb_sub ← floor(β_max·C_2·i / π)
           |   | for j ∈ 1..nb_sub
           |   |   bnd_sub_j ← j·π / C_2·i
           |   | bnds ← stack(bnds, bnd_sub)
           | bnds ← sort(bnds)
           | tol ← 0.0001·β_max
           | n ← 0
           | for i ∈ 1..(rows(bnds) − 1)
           |   if | bnds_(i−1) < bnds_i
           |      | f(bnds_i) ≠ 0
           |      | (f(bnds_(i−1))·f(bnds_i)) ≤ 0
           |     | UB ← bnds_(i−1)  if f(bnds_(i−1)) = 0
           |     | UB ← bnds_i  otherwise
           |     | LB ← bnds_(i−1)
           |     | while (UB − LB) ≥ tol
           |     |   | β_n ← LB + (UB − LB)/2
           |     |   | UB ← β_n  if f(UB)·f(β_n) ≥ 0
           |     |   | LB ← β_n  if f(LB)·f(β_n) ≥ 0
           |     | β_n ← UB
           |     | n ← n + 1
           | β
```

Plot of the transcendental function and calculated roots, i.e., system eigenvalues

$n := 0 .. (\text{rows}(\beta\_array) - 1)$

$\beta := \beta\_array$

$zero\_array_n := 0$

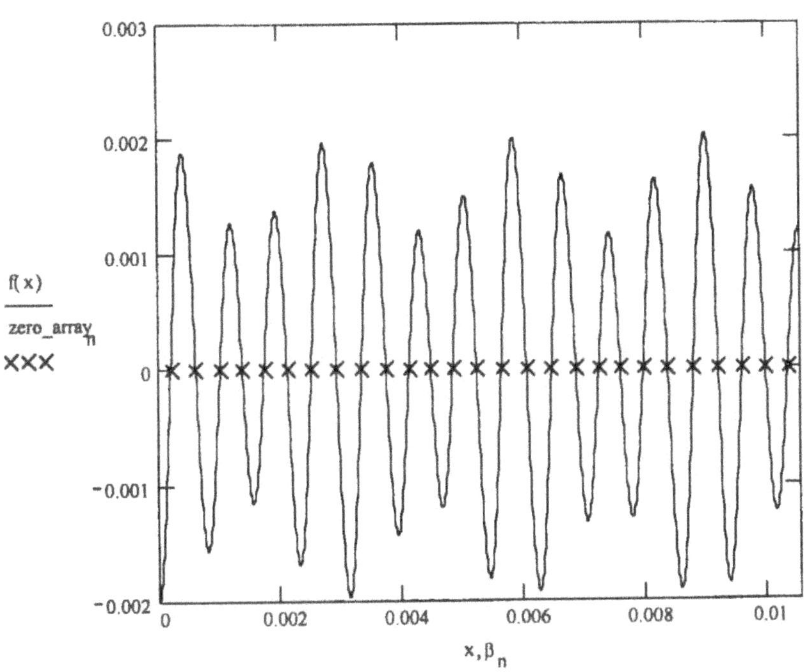

Calculate System Eigenfunctions

Define the eigenfunction coefficients, $A_i\,(\beta_n)$'s and $B_i\,(\beta_n)$'s

$$A(i,\beta) := \begin{cases} 0 & \text{if } i=1 \\ -\left(\dfrac{\gamma_1}{\gamma_2}\right)\cdot\cos\left(\beta\cdot\dfrac{a}{\sqrt{\alpha_2}}\right)\cdot\sin\left(\beta\cdot\dfrac{a}{\sqrt{\alpha_1}}\right) + \cos\left(\beta\cdot\dfrac{a}{\sqrt{\alpha_1}}\right)\cdot\sin\left(\beta\cdot\dfrac{a}{\sqrt{\alpha_2}}\right) & \text{otherwise} \end{cases}$$

$$B(i,\beta) := \begin{cases} 1 & \text{if } i=1 \\ \left(\dfrac{\gamma_1}{\gamma_2}\right)\cdot\sin\left(\beta\cdot\dfrac{a}{\sqrt{\alpha_2}}\right)\cdot\sin\left(\beta\cdot\dfrac{a}{\sqrt{\alpha_1}}\right) + \cos\left(\beta\cdot\dfrac{a}{\sqrt{\alpha_2}}\right)\cdot\cos\left(\beta\cdot\dfrac{a}{\sqrt{\alpha_1}}\right) & \text{otherwise} \end{cases}$$

Define eigenfunction, $\Psi_i(\beta_n)$

$$\Psi(i,z,\beta) := A(i,\beta)\cdot\sin\left(\sqrt{\frac{R_i}{D_i}}\cdot z\cdot\beta\right) + B(i,\beta)\cdot\cos\left(\sqrt{\frac{R_i}{D_i}}\cdot z\cdot\beta\right)$$

where:
'i' = layer number
'z' = position
'b' = an eigenvalue

Calculate the normalization integral, initialization integral, and differential of the surface eigenfunction

Normalization integral

$$\text{NormIntegral}(\beta) := R_1\cdot\int_0^a (\Psi(1,z,\beta))^2 dz + R_2\cdot\int_a^b (\Psi(2,z,\beta))^2 dz$$

$$N_n := \text{NormIntegral}(\beta_n)$$

Initialization integral

$$\text{layer}(z) := \begin{vmatrix} 2 \text{ if } a < z \le b \\ 1 \text{ otherwise} \end{vmatrix}$$

subroutine used to locate which layer a particular value of 'z' belongs to

$$\text{InitIntegral}(\beta) := \begin{vmatrix} \text{sum} \leftarrow 0 \\ \text{for } k \in 0..(\text{rows}(c0)-2) \\ \quad \begin{vmatrix} i \leftarrow \text{layer}(c0_{k,0}) \\ \text{sum} \leftarrow \text{sum} + c0_{k,1}\cdot R_i\cdot\int_{c0_{k+1,0}}^{c0_{k,0}} \Psi(i,z,\beta)\,dz \end{vmatrix} \\ \text{sum} \end{vmatrix}$$

subroutine used to calculate the initialization integral from the array 'c0'

$$I_n := \text{InitIntegral}(\beta_n)$$

Surface eigenfunction differential

$$d\Psi_n := \left(\sqrt{\frac{R_2}{D_2}}\cdot\beta_n\right)\cdot\left(A(2,\beta_n)\cdot\cos\left(\sqrt{\frac{R_2}{D_2}}\cdot b\cdot\beta_n\right) - B(2,\beta_n)\cdot\sin\left(\sqrt{\frac{R_2}{D_2}}\cdot b\cdot\beta_n\right)\right)$$

Calculate the final concentration profile

$z\_max\_div := 100$ $\qquad j := 0 .. z\_max\_div$ $\qquad z_j := b \cdot \left(\frac{j}{z\_max\_div}\right)$

$time_{max} := max(time)$

$$ConcCalc(z) := \left| \begin{array}{l} i \leftarrow layer(z) \\ \sum_n \left[e^{-(\beta_n)^2 \cdot time_{max}} \cdot \frac{1}{N_n} \cdot I_n \cdot \Psi(i, z, \beta_n) \right] \end{array} \right.$$

$c_j := ConcCalc(z_j)$

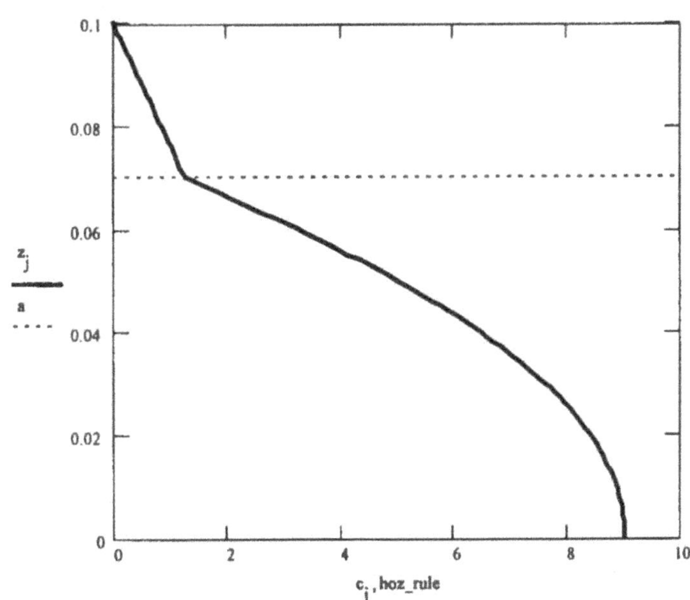

Hazardous Substance Research Center
Louisiana State University

PROJECT Three-layer diffusion model

DESCRIPTION (Model 6-2: concentration profile and surface flux)

Calculation of a concentration profile from diffusion in a three-layer composite
Ref: Contaminant Transport in Soils and Sediments, §6

CALCULATION BY Bruce Choy DATE 27 Sept 1997

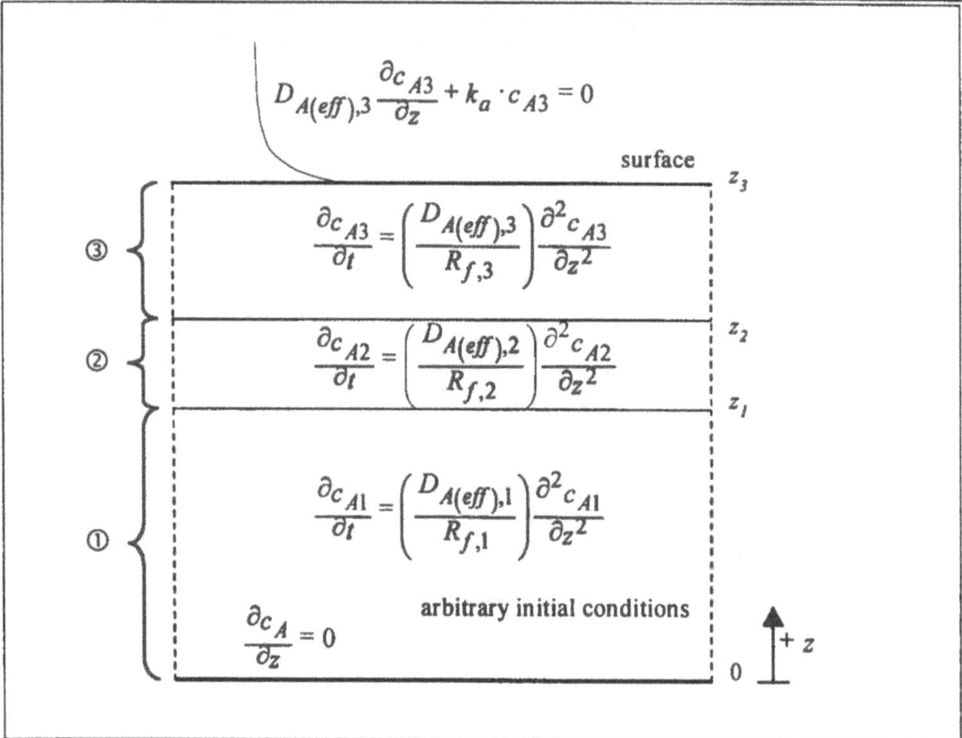

System parameters:

Diffusion coefficients, retardation factors, and layer locations as specified in the above diagram

$D_3 := 5\cdot10^{-7}$ m²/s	$R_3 := 600$	$z_3 := 0.100$ m	= surface layer
$D_2 := 1\cdot10^{-7}$	$R_2 := 2000$	$z_2 := 0.080$	= middle layer
$D_1 := 0.5\cdot10^{-7}$	$R_1 := 2000$	$z_1 := 0.070$	= bottom layer

$k_a := -0.001$ m/s

Simulation parameter:

$time := 100\cdot3600$ seconds note: initial concentration is specified below

Simulation parameter: $m\_max := 200$

$m := 1..\,m\_max$ = array of time steps in which surface flux profile is calculated from

$time_m := (10\cdot m + 10)\cdot 3600$ seconds

$time_{m\_max} = 7.236\cdot 10^6$

Calculation of the coefficients for eigenfunction and eigenvalues given the specified system configuration:

$i := 1..\,3$ $\qquad \alpha_i := \dfrac{D_i}{R_i}$ $\qquad \gamma_i := \sqrt{D_i \cdot R_i}$ $\qquad \eta_i := \dfrac{z_i}{\sqrt{\alpha_i}}$

$\phi_1 := \dfrac{z_3 - z_2}{\sqrt{\alpha_3}}$ $\qquad \phi_2 := \dfrac{z_2 - z_1}{\sqrt{\alpha_2}}$

$$
\begin{aligned}
A(i,\beta) := \; & x \leftarrow 0 \text{ if } i = 1\\
& denom \leftarrow \gamma_3 \cdot \beta \cdot \left(\gamma_2 \cdot \cos(\beta \cdot \phi_1) \cdot \cos(\beta \cdot \phi_2) - \gamma_3 \cdot \sin(\beta \cdot \phi_1) \cdot \sin(\beta \cdot \phi_2)\right)\\
& denom \leftarrow denom + k_a \cdot \left(\gamma_2 \cdot \sin(\beta \cdot \phi_1) \cdot \cos(\beta \cdot \phi_2) + \gamma_3 \cdot \cos(\beta \cdot \phi_1) \cdot \sin(\beta \cdot \phi_2)\right)\\
& \text{if } i = 2\\
& \quad numer \leftarrow \gamma_3 \cdot \beta \cdot \cos(\beta \cdot \eta_1) \cdot \left(\gamma_2 \cdot \sin(\beta \cdot \eta_2) \cdot \cos(\beta \cdot \phi_1) + \gamma_3 \cdot \cos(\beta \cdot \eta_2) \cdot \sin(\beta \cdot \phi_1\right.\\
& \quad numer2 \leftarrow k_a \cdot \cos(\beta \cdot \eta_1) \cdot \left(\gamma_2 \cdot \sin(\beta \cdot \eta_2) \cdot \sin(\beta \cdot \phi_1) - \gamma_3 \cdot \cos(\beta \cdot \eta_2) \cdot \cos(\beta \cdot \phi_1\right.\\
& \quad numer \leftarrow numer + numer2\\
& \quad x \leftarrow \frac{numer}{denom}\\
& \text{otherwise}\\
& \quad numer \leftarrow \gamma_2 \cdot \gamma_3 \cdot \beta \cdot \cos(\beta \cdot \eta_1) \cdot \sin(\beta \cdot \eta_3) - k_a \cdot \gamma_2 \cdot \cos(\beta \cdot \eta_1) \cdot \cos(\beta \cdot \eta_3)\\
& \quad x \leftarrow \frac{numer}{denom}\\
& x \leftarrow 0 \text{ if } i = 1\\
& x
\end{aligned}
$$

$$
B(i,\beta) := \left|
\begin{array}{l}
x \leftarrow 1 \ \text{ if } i=1 \\
denom \leftarrow \gamma_3 \cdot \beta \cdot \left(\gamma_2 \cdot \cos(\beta \cdot \phi_1) \cdot \cos(\beta \cdot \phi_2) - \gamma_3 \cdot \sin(\beta \cdot \phi_1) \cdot \sin(\beta \cdot \phi_2)\right) \\
denom \leftarrow denom + k_a \cdot \left(\gamma_2 \cdot \sin(\beta \cdot \phi_1) \cdot \cos(\beta \cdot \phi_2) + \gamma_3 \cdot \cos(\beta \cdot \phi_1) \cdot \sin(\beta \cdot \phi_2)\right) \\
\text{if } i=2 \\
\quad \left|
\begin{array}{l}
numer \leftarrow \gamma_3 \cdot \beta \cdot \cos(\beta \cdot \eta_1) \cdot \left(\gamma_2 \cdot \cos(\beta \cdot \eta_2) \cdot \cos(\beta \cdot \phi_1) - \gamma_3 \cdot \sin(\beta \cdot \eta_2) \cdot \sin(\beta \cdot \phi_1)\right) \\
numer2 \leftarrow k_a \cdot \cos(\beta \cdot \eta_1) \cdot \left(\gamma_2 \cdot \cos(\beta \cdot \eta_2) \cdot \sin(\beta \cdot \phi_1) + \gamma_3 \cdot \sin(\beta \cdot \eta_2) \cdot \cos(\beta \cdot \phi_1)\right) \\
numer \leftarrow numer + numer2 \\
x \leftarrow \dfrac{numer}{denom}
\end{array}
\right. \\
\text{if } i=3 \\
\quad \left|
\begin{array}{l}
numer \leftarrow \gamma_2 \cdot \gamma_3 \cdot \beta \cdot \cos(\beta \cdot \eta_1) \cdot \cos(\beta \cdot \eta_3) + k_a \cdot \gamma_2 \cdot \cos(\beta \cdot \eta_1) \cdot \sin(\beta \cdot \eta_3) \\
x \leftarrow \dfrac{numer}{denom}
\end{array}
\right. \\
x
\end{array}
\right.
$$

Transcendental function definition

$$C_1 := k_a \cdot \left[(\gamma_1 \cdot \gamma_2) + (\gamma_1 \cdot \gamma_3) + (\gamma_2)^2 + (\gamma_2 \cdot \gamma_3)\right] \qquad C_2 := \phi_1 + \phi_2 + \eta_1$$

$$C_3 := k_a \cdot \left[-(\gamma_1 \cdot \gamma_2) - (\gamma_1 \cdot \gamma_3) + (\gamma_2)^2 + (\gamma_2 \cdot \gamma_3)\right] \qquad C_4 := \phi_1 + \phi_2 - \eta_1$$

$$C_5 := k_a \cdot \left[(\gamma_1 \cdot \gamma_2) - (\gamma_1 \cdot \gamma_3) - (\gamma_2)^2 + (\gamma_2 \cdot \gamma_3)\right] \qquad C_6 := \phi_1 - \phi_2 + \eta_1$$

$$C_7 := k_a \cdot \left[-(\gamma_1 \cdot \gamma_2) + (\gamma_1 \cdot \gamma_3) - (\gamma_2)^2 + \gamma_2 \cdot \gamma_3\right] \qquad C_8 := \phi_1 - \phi_2 - \eta_1$$

$$C_9 := \gamma_3 \cdot \left[-(\gamma_1 \cdot \gamma_2) - (\gamma_1 \cdot \gamma_3) - (\gamma_2)^2 - (\gamma_2 \cdot \gamma_3)\right] \qquad C_{10} := C_2$$

$$C_{11} := \gamma_3 \cdot \left[(\gamma_1 \cdot \gamma_2) + (\gamma_1 \cdot \gamma_3) - (\gamma_2)^2 - (\gamma_2 \cdot \gamma_3)\right] \qquad C_{12} := C_4$$

$$C_{13} := \gamma_3 \cdot \left[-(\gamma_1 \cdot \gamma_2) + (\gamma_1 \cdot \gamma_3) + (\gamma_2)^2 - (\gamma_2 \cdot \gamma_3)\right] \qquad C_{14} := C_6$$

$$C_{15} := \gamma_3 \cdot \left[(\gamma_1 \cdot \gamma_2) - (\gamma_1 \cdot \gamma_3) + (\gamma_2)^2 - (\gamma_2 \cdot \gamma_3)\right] \qquad C_{16} := C_8$$

$$
\begin{aligned}
f(\beta) := {} & C_1 \cdot \cos(C_2 \cdot \beta) + C_3 \cdot \cos(C_4 \cdot \beta) + C_5 \cdot \cos(C_6 \cdot \beta) + C_7 \cdot \cos(C_8 \cdot \beta) \ \ldots \\
& + C_9 \cdot \beta \cdot \sin(C_{10} \cdot \beta) + C_{11} \cdot \beta \cdot \sin(C_{12} \cdot \beta) + C_{13} \cdot \beta \cdot \sin(C_{14} \cdot \beta) + C_{15} \cdot \beta \cdot \sin(C_{16} \cdot \beta)
\end{aligned}
$$

Array of significant eigenvalues, β

$\omega := 10 \cdot 10^{-10}$ ... time constant for calculation

$$\beta\_max := \omega \cdot \frac{1}{\sqrt{time_{m\_max}}} \cdot max\left[\begin{bmatrix} (\alpha_1)^{-1} \\ (\alpha_2)^{-1} \\ (\alpha_3)^{-1} \end{bmatrix}\right]$$

```
β_array := | bnds_0 ← 0
           | for i ∈ 1..4
           |   | nb_sub ← floor( β_max·C_{2,i} / π )      ... range of significance β=[0,β_max)
           |   | for j ∈ 1..nb_sub
           |   |   bnd_sub_j ← j·π / C_{2,i}
           |   | bnds ← stack(bnds, bnd_sub)
           |   | for j ∈ 0..nb_sub − 1
           |   |   bnd_sub_j ← (j + 0.5)·π / C_{2,i}
           |   | bnds ← stack(bnds, bnd_sub)
           | bnds ← sort(bnds)
           | tol ← 0.0001·β_max
           | n ← 0
           | for i ∈ 1..(rows(bnds) − 1)
           |   if | bnds_{i−1} < bnds_i
           |      | f(bnds_i) ≠ 0
           |      | (f(bnds_{i−1})·f(bnds_i)) ≤ 0
           |     | UB ← bnds_{i−1}  if f(bnds_{i−1}) = 0
           |     | UB ← bnds_i  otherwise
           |     | LB ← bnds_{i−1}
           |     | while (UB − LB) ≥ tol
           |     |   | β_n ← LB + (UB − LB)/2
           |     |   | UB ← β_n  if f(UB)·f(β_n) ≥ 0
           |     |   | LB ← β_n  if f(LB)·f(β_n) ≥ 0
           |     | β_n ← UB
           |     | n ← n + 1
           | β
```

notes:

1. The adjustable parameter ω is used to determine the range of significance (i.e., the range in which the eigenvalues are calculated).
2. Choose an ω such that the final concentration profile has a high enough resolution frequency to yield an appropriate (smooth) curve.
3. Usually 10-20 eigenvalues, i.e., 'β's is a reasonable number for most 'time' variables.
4. Higher number of eigenvalues are required for short times (i.e., higher values of ω).
5. The number of eigenvalues calculated by the rigorous bracketing-bisection algorithm 'β-array', can be seen in the graph on the following page.

Plot of the transcendental function and calculated roots

$n := 0..(\text{rows}(\beta\_array) - 1)$

$\beta := \beta\_array$

$zero\_array_n := 0$

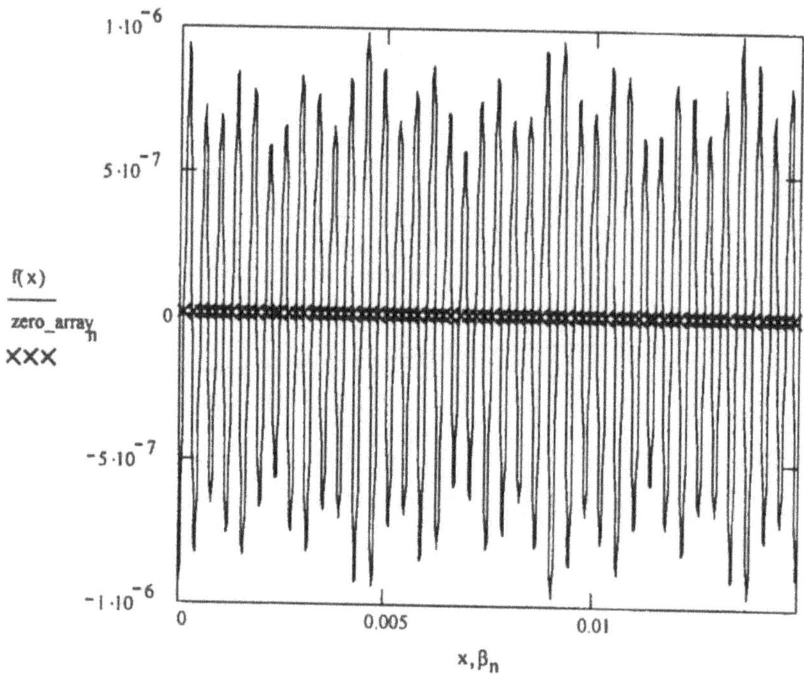

Initial concentration specification:

$$c0 := \begin{bmatrix} 0.10 & 0 \\ 0.08 & 0 \\ 0.07 & 10 \\ 0.02 & 10 \\ 0 & 0 \end{bmatrix}$$

notes:

1. The c0 array should be in order of descending depth where the left entry is 'z' and the right is 'conc' of the layer directly below it.
2. Specify the interface positions explicitly in the array (as this simplifies the integration process).
3. Top entry must contain the value of z_3 in the 'z' position and the concentration of the surface zone in the 'conc' position.
4. Bottom entry must contain a zero-zero ordered pair.

```
plot_data := | x ← [c0_{0,0}  0]
             | for i ∈ 0.. (rows(c0) − 2)
             |    | y_{0,0} ← c0_{i,0}
             |    | y_{0,1} ← c0_{i,1}
             |    | y_{1,0} ← c0_{i+1,0}
             |    | y_{1,1} ← c0_{i,1}
             |    | x ← stack(x, y)
             | x
```

$p := 0..\,(\mathrm{rows}(plot\_data) - 1)$

$InitConc_p := plot\_data_{p,1}$

$depth_p := plot\_data_{p,0}$

Plot of initial concentration

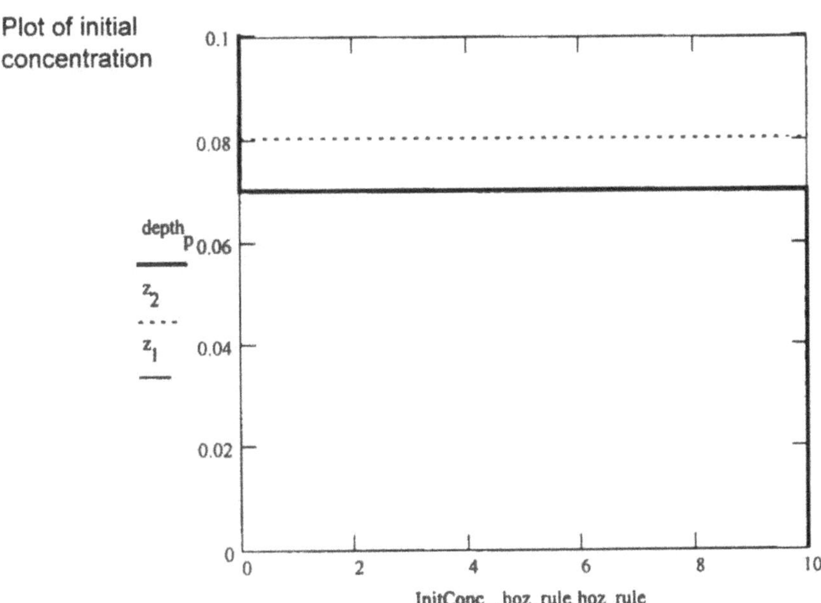

Define eigenfunction

$$\Psi(i,b,\zeta) := A(i,b)\cdot\sin\left(\sqrt{\frac{R_i}{D_i}}\cdot b\cdot\zeta\right) + B(i,b)\cdot\cos\left(\sqrt{\frac{R_i}{D_i}}\cdot b\cdot\zeta\right)$$

Define normalization integral

$$z_0 := 0$$

$$NormIntegral(b) := \sum_i R_i\cdot\int_{z_{i-1}}^{z_i} (\Psi(i,b,\zeta))^2\,d\zeta$$

$$N_n := NormIntegral(\beta_n)$$

Define the initialization integral

$$\mathrm{layer}(\zeta) := \begin{cases} 3 & \text{if } z_2 < \zeta \le z_3 \\ 2 & \text{if } z_1 < \zeta \le z_2 \\ 1 & \text{otherwise} \end{cases}$$

$$\mathrm{InitIntegral}(b) := \left| \begin{array}{l} x \leftarrow 0 \\ \text{for } k \in 0..\left(\mathrm{rows}(c0) - 2\right) \\ \quad \left| \begin{array}{l} i \leftarrow \mathrm{layer}\left(c0_{k,0}\right) \\ x \leftarrow x + c0_{k,1} \cdot R_i \cdot \int_{c0_{k+1,0}}^{c0_{k,0}} \Psi(i,b,\zeta)\,d\zeta \end{array} \right. \\ x \end{array} \right.$$

$$I_n := \mathrm{InitIntegral}\left(\beta_n\right)$$

Calculate the concentration profile

$$z\_max\_div := 100 \qquad j := 0..\, z\_max\_div \qquad \zeta_j := z_3 \cdot \left(\frac{j}{z\_max\_div}\right)$$

$$\mathrm{ConcCalc}(x) := \left| \begin{array}{l} i \leftarrow \mathrm{layer}(x) \\ \sum_n \left[e^{-\left(\beta_n\right)^2 \cdot time_{m\_max}} \cdot \frac{1}{N_n} \cdot I_n \cdot \Psi\left(i, \beta_n, x\right) \right] \end{array} \right.$$

$$c_j := \mathrm{ConcCalc}\left(\zeta_j\right)$$

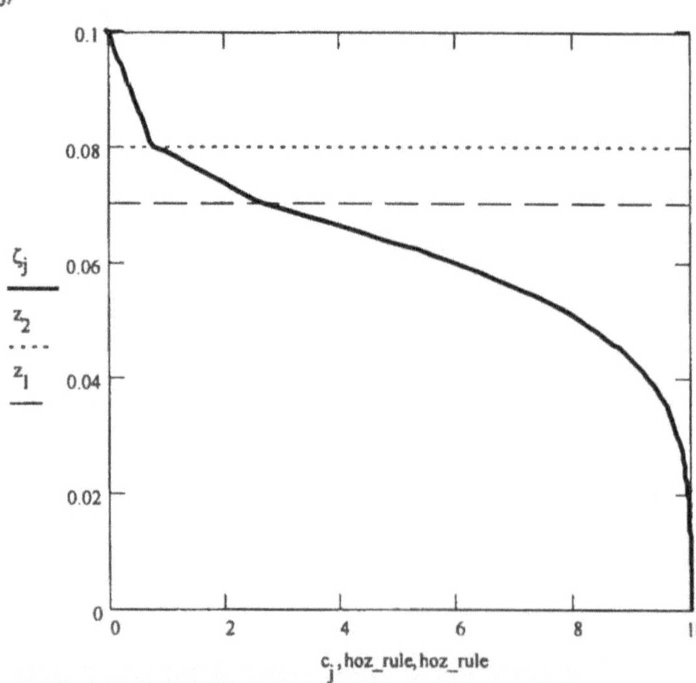

Calculate surface flux

$$d\Psi_n := \left(\sqrt{\frac{R_3}{D_3}}\cdot\beta_n\right)\cdot\left(A(3,\beta_n)\cdot\cos\left(\sqrt{\frac{R_3}{D_3}}\cdot z_3\cdot\beta_n\right) - B(3,\beta_n)\cdot\sin\left(\sqrt{\frac{R_3}{D_3}}\cdot z_3\cdot\beta_n\right)\right)$$

$$SurfaceFlux_m := -D_3\cdot\left[\sum_n\left[e^{-(\beta_n)^2\cdot time_m}\cdot\frac{1}{N_n}\cdot I_n\cdot d\Psi_n\right]\right]$$

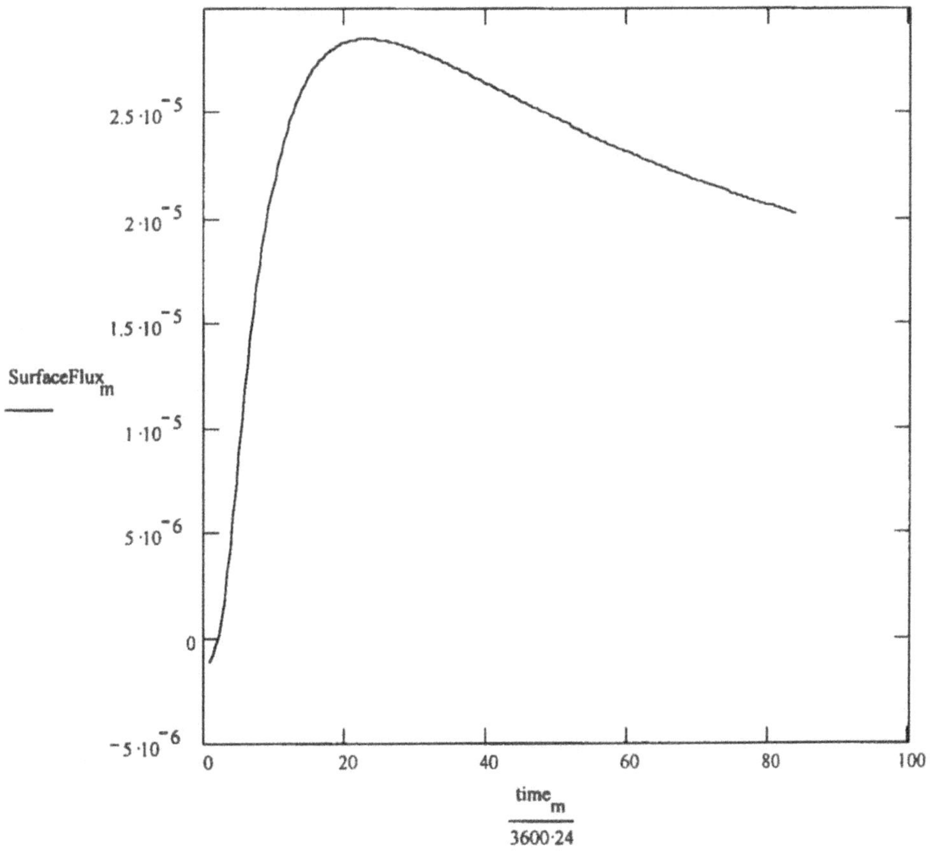